T0245236

CAMBRIDGE LIBRARY COLLECTION

Books of enduring scholarly value

Darwin

Two hundred years after his birth and 150 years after the publication of 'On the Origin of Species', Charles Darwin and his theories are still the focus of worldwide attention. This series offers not only works by Darwin, but also the writings of his mentors in Cambridge and elsewhere, and a survey of the impassioned scientific, philosophical and theological debates sparked by his 'dangerous idea'.

On the Various Contrivances by which British and Foreign Orchids are Fertilised by Insects

In this investigation of orchids, first published in 1862, Darwin expands on a point made in *On the Origin of Species* that he felt required further explanation, namely that he believes it to be 'a universal law of nature that organic beings require an occasional cross with another individual'. Darwin explains the method by which orchids are fertilised by insects, and argues that the intricate structure of their flowers evolved to favour cross pollination because of its advantages to the species. The book is written in Darwin's usual precise and elegant style, accessible despite its intricate detail. It includes a brief explanation of botanical terms and is illustrated with 34 woodcuts.

Cambridge University Press has long been a pioneer in the reissuing of out-of-print titles from its own backlist, producing digital reprints of books that are still sought after by scholars and students but could not be reprinted economically using traditional technology. The Cambridge Library Collection extends this activity to a wider range of books which are still of importance to researchers and professionals, either for the source material they contain, or as landmarks in the history of their academic discipline.

Drawing from the world-renowned collections in the Cambridge University Library, and guided by the advice of experts in each subject area, Cambridge University Press is using state-of-the-art scanning machines in its own Printing House to capture the content of each book selected for inclusion. The files are processed to give a consistently clear, crisp image, and the books finished to the high quality standard for which the Press is recognised around the world. The latest print-on-demand technology ensures that the books will remain available indefinitely, and that orders for single or multiple copies can quickly be supplied.

The Cambridge Library Collection will bring back to life books of enduring scholarly value (including out-of-copyright works originally issued by other publishers) across a wide range of disciplines in the humanities and social sciences and in science and technology.

On the Various Contrivances by which British and Foreign Orchids are Fertilised by Insects

And On the Good Effect of Intercrossing

CHARLES DARWIN

CAMBRIDGE
UNIVERSITY PRESS

CAMBRIDGE UNIVERSITY PRESS

Cambridge, New York, Melbourne, Madrid, Cape Town, Singapore,
São Paolo, Delhi, Dubai, Tokyo, Mexico City

Published in the United States of America by Cambridge University Press, New York

www.cambridge.org
Information on this title: www.cambridge.org/9781108027151

© in this compilation Cambridge University Press 2011

This edition first published 1862
This digitally printed version 2011

ISBN 978-1-108-02715-1 Paperback

ON

THE VARIOUS CONTRIVANCES

BY WHICH

BRITISH AND FOREIGN ORCHIDS

ARE

FERTILISED BY INSECTS,

AND ON THE GOOD EFFECTS OF INTERCROSSING.

By CHARLES DARWIN, M.A., F.R.S., &c.

WITH ILLUSTRATIONS.

LONDON:

JOHN MURRAY, ALBEMARLE STREET.

1862.

LONDON : PRINTED BY W. CLOWES AND SONS, STAMFORD STREET,
AND CHARING CROSS.

CONTENTS.

——o——

CHAPTER V.

CHAPTER VI.

CHAPTER VII.

(v)

LIST OF WOODCUTS.

——o——

b

P.S.—I am much indebted to Mr. G. B. Sowerby for the pains which he has taken in making the Diagrams as intelligible as possible.

INSTRUCTION TO BINDER.

Let the Woodcut No. I. face page 18.

ON THE

FERTILISATION OF ORCHIDS BY INSECTS,

&c. &c.

INTRODUCTION.

THE object of the following work is to show
that the contrivances by which Orchids are
fertilised, are as varied and almost as perfect
as any of the most beautiful adaptations in the
animal kingdom; and, secondly, to show that
these contrivances have for their main object
the fertilisation of each flower by the pollen
of another flower. In my volume ' On the
Origin of Species ' I have given only general
reasons for my belief that it is apparently a
universal law of nature that organic beings
require an occasional cross with another individual; or, which is almost the same thing, that
no hermaphrodite fertilises itself for a perpetuity of generations. Having been blamed
for propounding this doctrine without giving

B

ample facts, for which I had not, in that work, sufficient space, I wish to show that I have not spoken without having gone into details.

I have been led to publish this little treatise separately, as it has become inconveniently large to be incorporated with the rest of the discussion on the same subject. And I have thought, that, as Orchids are universally acknowledged to rank amongst the most singular and most modified forms in the vegetable kingdom, the facts to be presently given might lead some observers to look more curiously into the habits of our several native species. An examination of their many beautiful contrivances will exalt the whole vegetable kingdom in most persons' estimation. I fear, however, that the necessary details will be too minute and complex for any one who has not a strong taste for Natural History. This treatise affords me also an opportunity of attempting to show that the study of organic beings may be as interesting to an observer who is fully convinced that the structure of each is due to secondary laws, as to one who views every trifling detail of structure as the result of the direct interposition of the Creator.

I must premise that Christian Konrad Sprengel, in his curious and valuable work, ' Das entdeckte Geheimniss der Natur,' published in 1793, gave an excellent outline of the action of the several parts in Orchids; for he well knew the position of the stigma; and he discovered that insects were necessary to remove the pollen-masses, by pushing open the pouch and coming into contact with the enclosed sticky glands. But he overlooked many curious contrivances,—a consequence, apparently, of his belief that the stigma generally receives the pollen of the same flower. Sprengel, likewise, has partially described the structure of Epipactis; but in the case of Listera he entirely misunderstood the remarkable phenomena characteristic of that genus, which has been so well described by Dr. Hooker in the ' Philosophical Transactions' for 1854. Dr. Hooker has given a full and accurate account, with drawings, of the structure of the parts, and of what takes place; but from not having attended to the agency of insects, he has not fully understood the object gained. Robert Brown,* in

* Linnæan Transactions,' 1833, vol. xvi. p. 704.

his celebrated paper in the 'Linnæan Transactions,' expresses his belief that insects are necessary for the fructification of most Orchids; but adds, that the fact of all the capsules on a dense spike not infrequently producing seed, seems hardly reconcileable with this belief: we shall hereafter see that this doubt is groundless. Many other authors have given facts and expressed their belief, more or less fully, on the necessity of insect-agency in the fertilisation of Orchids.

In the course of the following work I shall have the pleasure of expressing my deep obligation to several gentlemen for their unremitting kindness in sending me fresh specimens, without which aid this work would have been impossible. The trouble which several of my kind assistants have taken has been extraordinary: I have never once expressed a wish for aid or for information which has not been granted, as far as possible, in the most liberal spirit.

EXPLANATION OF TERMS.

In case any one should look at this treatise who has never attended to Botany, it may be convenient to explain the meaning of the common terms used. In most flowers the stamens, or male organs, surround in a ring the one or more female organs, called the pistils. In all common Orchids there is only one stamen, and this is confluent with the pistil forming the *Column*. The stamens consist of a filament, or supporting thread (rarely seen in British Orchids), which carries the anther; and within the anther the pollen, the male vivifying element, is included. The anther is divided into two cells, which are very distinct in most Orchids, so much so as to appear in some species like two separate anthers. The pollen in all common plants consists of fine granular powder : but in most Orchids the grains cohere in masses, which are often supported by a very curious appendage, called the *Caudicle*; as will hereafter be more fully explained. The pollen-masses, with their caudicles and other appendages, are called the *Pollinia*.

There are properly in Orchids three united
pistils, or female organs. The upper part of
the pistil has its anterior surface soft and
viscid, which forms the stigma. The two
lower stigmas are often completely confluent,
so as to appear as one. The stigma in the
act of fertilisation is penetrated by long tubes
emitted by the pollen-grains, which carry the
contents of the grains down to the ovules, or
young seeds in the ovarium.

Of the three pistils, which ought to be
present, the stigma of the upper one has been
modified into an extraordinary organ, called
the *Rostellum*, which in many Orchids presents no resemblance to a true stigma. The
rostellum either includes or is formed of
viscid matter; and in very many Orchids
the pollen-masses are firmly attached to a
portion of its exterior membrane, which is
removed, together with the pollen-masses, by
insects. This removeable portion consists in
most British Orchids of a small piece of
membrane, with a layer or ball of viscid
matter underneath, and I shall call it the
"*viscid disc*;" but in many exotic Orchids the
portion removed is so large and so import-

ant, that one part must be called, as before,
the viscid disc, and the other part the *pedicel*
of the rostellum, to the end of which pedicel
the pollen-masses are attached. Authors have
called that portion of the rostellum which is
removed the " gland," or the " retinaculum,"
from its apparent function of retaining the
pollen-masses in place. The pedicel, or pro-
longation of the rostellum, to which in many
exotic Orchids the pollen-masses are attached,
seems generally to have been confounded,
under the name of caudicle, with the true
caudicle of the pollen-masses, though their
nature and origin are totally different. The
part of the rostellum which is not removed,
and which includes the viscid matter, is sometimes called the " bursicula," or " fovea," or
" pouch." But it will be found most convenient to avoid all these terms, and to call the
whole modified stigma the rostellum—sometimes adding an adjective to define its shape;
and to call that portion of the rostellum which
is attached to and removed with the pollen-
masses the *viscid disc*, together in some cases
with its *pedicel*.

Lastly, the three outer divisions of the

flower are called *Sepals,* and form the calyx;
but, instead of being green, as in most common
flowers, they are generally coloured, like the
three inner divisions or *Petals* of the flower.
The one petal which commonly stands lowest
is larger than the others, and often assumes
most singular shapes; it is called the lower
lip, or *Labellum.* It secretes nectar, in order
to attract insects, and is often produced into a
long spur-like nectary.

(9)

CHAPTER I.

Structure of Orchis — Power of movement of the pollinia — Perfect adaptation of the parts in Orchis pyramidalis — On the insects which visit Orchids, and on the frequency of their visits — On the fertility and sterility of several Orchids — On the secretion of nectar, and on moths being purposely delayed in obtaining it.

FOR my purpose British Orchids may be divided into three groups, and the arrangement is, for the most part, a natural one. But I leave out of consideration the British species of Cypripedium, with its two anthers, of which I know nothing. Of these three groups the first consists of the Ophreæ, which have pollinia furnished at their lower ends with a caudicle, congenitally attached to a viscid disc. The anther stands above the rostellum. The Ophreæ include most of our common Orchids.

First, for the genus Orchis. The reader may find the following details rather difficult to understand; but I can assure him, if he will have patience to make out this first case, the succeeding cases will be easily intelligible.

B 3

The accompanying diagrams (Fig. I., p. 18) show the relative position of the more important organs in the flower of the Early Orchis (O. mascula). The sepals and the petals have been removed, excepting the labellum with its nectary. The nectary is shown only in the side view (*n* Fig. A); for its enlarged orifice is almost hidden in shade in (B) the front view. The stigma (*s*) is bilobed, and consists of two almost confluent stigmas; it lies under the pouch-formed (*r*) rostellum. The anther (*a* in B and A) consists of two rather widely separated cells, which are longitudinally open in front : each cell includes a pollen-mass or pollinium.

A pollinium removed out of one of the two anther-cells is represented by Fig. C ; it consists of a number of wedge-formed packets of pollen-grains (see Fig. F, in which the packets are forcibly separated), united together by excessively elastic, thin threads. These threads become confluent at the lower end of each pollen-mass, and compose the (*c* C) straight elastic caudicle. The end of the caudicle is firmly attached to the viscid disc (*d* C), which consists (as may be seen in the section, Fig. E) of a minute oval piece of membrane, with a ball of viscid matter on its under side. Each pollinium has its separate disc; and the two balls of viscid matter lie enclosed together (Fig. D) within the rostellum.

The rostellum is a nearly spherical, somewhat pointed projection (*r* Figs. A and B) overhanging the two almost confluent stigmas, and must be fully described, as every detail of its structure is full of signification. A section through one of the discs and balls of viscid matter is given (Fig. E); and a front view of both viscid discs within the rostellum (Fig. D) is likewise given. This latter figure (D) probably best serves to explain the structure of the rostellum; but it must be understood that the front lip is here considerably depressed. The lowest part of the anther is united to the back of the rostellum, as may be seen in Fig. B. At an early period of growth the rostellum consists of a mass of polygonal cells, full of brownish matter, which cells soon resolve themselves into two balls of an extremely viscid semi-fluid substance, void of structure. These viscid masses are slightly elongated, almost flat on the top, and convex below. They lie quite free within the rostellum (being surrounded by fluid), except at the back, where each viscid ball firmly adheres to a small portion or disc of the exterior membrane of the rostellum. The ends of the two caudicles are strongly attached to these two little discs of membrane.

The membrane forming the whole exterior surface of the rostellum is at first continuous; but as soon as the flower opens the slightest touch causes it to rupture transversely in a sinuous line, in front of the anther-cells and of the little crest or fold of membrane (see Fig. D) between them. This act of rupturing makes no difference in the shape of the rostellum, but converts the front part into a lip, which can easily be depressed. This lip is represented considerably depressed in Fig. D, and its edge is seen, Fig. B, in the front view. When the lip is thoroughly depressed, the two balls of viscid matter are exposed. Owing to the elasticity of the hinder part or hinge, the lip or pouch, when not pressed down, springs up and again encloses the two viscid balls.

I will not affirm that the rupturing of the exterior membrane of the rostellum never takes place spontaneously; and no doubt the membrane is prepared for the rupture by having become very weak along defined lines ; but several times I saw the act ensue from an excessively slight touch—so slight that I conclude that the action is not simply mechanical, but, for the want of a better term, may be called vital. We shall hereafter meet with other cases, in which the slightest touch or the vapour of chloroform causes the exterior membrane of the rostellum to rupture along certain defined lines.

At the same time that the rostellum becomes transversely ruptured in front, it probably (for it was impossible to ascertain this fact from the position of the parts) ruptures behind in two oval lines, thus separating and freeing from the rest of the exterior surface of the rostellum the two little discs of membrane, to which externally the two caudicles are attached, and to which internally the two balls of viscid matter adhere. The line of rupture is thus very complex, but strictly defined.

As the two anther-cells open longitudinally in front from top to bottom, even before the flower expands, as soon as the rostellum is

properly ruptured from the effects of a slight
touch, its lip can be easily depressed, and, the
two little discs of membrane being already
separate, the two pollinia now lie absolutely
free, but are still embedded in their proper
places. So that the packets of pollen and the
caudicles lie within the anther-cells; the discs
still form part of the posterior surface of the
rostellum, but are separate; and the balls of
viscid matter still lie concealed within the
rostellum.

Now let us see how this complex mechanism
acts. Let us suppose an insect to alight on
the labellum, which forms a good landing-
place, and to push its head into the chamber
(see side view Fig. I., A, or front view, B) at
the back of which lies the stigma (s), in order
to reach with its proboscis the end of the nectary; or, which does equally well to show the
action, push a sharply-pointed common pencil
into the nectary. Owing to the pouch-formed
rostellum projecting into the gangway of the
nectary, it is scarcely possible that any object
can be pushed into it without the rostellum
being touched. The exterior membrane of the
rostellum then ruptures in the proper lines,

and the lip or pouch is most easily depressed. When this is effected, one or both of the viscid balls will almost infallibly touch the intruding body.   So viscid are these balls that whatever they touch they firmly stick to.   Moreover the viscid matter has the peculiar chemical quality of setting, like a cement, hard and dry in a few minutes' time.   As the anther-cells are open in front, when the insect withdraws its head, or when the pencil is withdrawn, one pollinium, or both, will be withdrawn, firmly cemented to the object, projecting up like horns, as shown (Fig. II.) by the upper figure.

Fig. II.

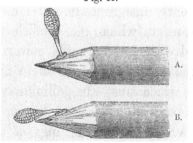

A. Pollen-mass of O. mascula, when first attached.
B.   do.   do.   after the act of depression.

The firmness of the attachment of the cement is very necessary, as we shall immediately see ; for if the pollinia were to fall sideways or

backwards they could never fertilise the flower.
From the position in which the two pollinia
lie in their cells, they diverge a little when
attached to any object. Now let us suppose
our insect to fly to another flower, or insert
the pencil (A, Fig. II.), with the attached pollinium, into the same or into another nectary :
by looking at the diagram (Fig. I., A) it will
be evident that the firmly attached pollinium
will be simply pushed against or into its old
position, namely, into its anther-cell. How
then can the flower be fertilised? This is
effected by a beautiful contrivance : though
the viscid surface remains immoveably affixed,
the apparently insignificant and minute disc
of membrane to which the caudicle adheres
is endowed with a remarkable power of contraction (as will hereafter be more minutely
described), which causes the pollinium to sweep
through about 90 degrees, always in one
direction, viz., towards the apex of the proboscis or pencil, in the course, on an average,
of thirty seconds. The position of the pollinium after the movement is shown at B in
Fig. II. Now after this movement and
interval of time (which would allow the insect

to fly to another flower), it will be seen, by turning to the diagram (Fig. I., A), that, if the pencil be inserted into the nectary, the thick end of the pollinium will exactly strike the stigmatic surface.

Here again comes into play another pretty adaptation, long ago noticed by Robert Brown.* The stigma is very viscid, but not so viscid as when touched to pull the whole pollinium off the insect's head or off the pencil, yet sufficiently viscid to break the elastic threads (Fig. I., F) by which the packets of pollen-grains are tied together, and leave some of them on the stigma. Hence a pollinium attached to an insect or to the pencil can be applied to many stigmas, and will fertilise all. I have seen the pollinia of Orchis pyramidalis adhering to the proboscis of a moth, with the stump-like caudicle alone left, all the packets of pollen having been left glued to the stigmas of the flowers successively visited.

One or two little points must still be noticed. The balls of viscid matter within the pouch-formed rostellum are surrounded with fluid; and this is very important, for,

* ' Transactions of the Linnæan Society,' vol. xvi. p. 731.

as already mentioned, the viscid matter sets
hard when exposed to the air for a very
short time. I have pulled the balls out of
their pouches and have found that in a few
minutes they entirely lost their power of adhesion. Again, the little discs of membrane,
the movement of which, as causing the movement of the pollinium, is so absolutely indispensable for the fertilisation of the flower, lie
at the upper and back surface of the rostellum,
and are closely enfolded and thus kept damp
within the bases of the anther-cells; and this is
very necessary, as an exposure of about thirty
seconds causes the movement of depression to
take place; but as long as the disc is kept
damp the pollinium remains ready for action
whenever removed by an insect.

Lastly, as I have shown, the pouch, after
being depressed, springs up to its former position; and this is of great service; for if this
action did not take place, and an insect after
depressing the lip failed to remove either viscid
ball, or if it removed one alone, in the first case
both, and in the second case one of the viscid
balls would be left exposed to the air; consequently they would quickly lose all adhesive-

Fig. I.

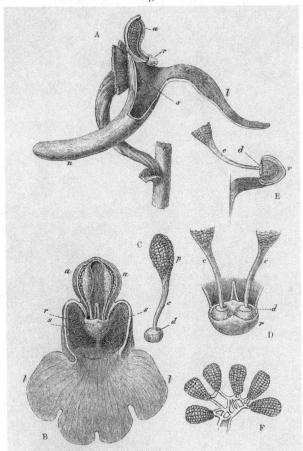

ORCHIS MASCULA.

Description of Fig. I.

a. anther.   *n.* nectary.
r. rostellum.   p. pollinium or pollen-mass.
s. stigma.   c. caudicle of pollinium.
l. labellum.   d. viscid disc of pollinium.

A. Side view of flower, with all the petals and sepals cut off except the labellum, of which the near half is cut away, as well as the upper portion of the near side of the nectary.
B. Front view of flower, with all sepals and petals removed, except the labellum.
C. One pollinium or pollen-mass, showing the packets of pollen-grains, the caudicle, and viscid disc.
D. Front view of the discs and caudicles of both pollinia within the rostellum, with its lip depressed.
E. Section through one side of the rostellum, with the included disc and caudicle of one pollinium.
F. Packets of pollen-grains, tied together by elastic threads, here extended. (Copied from Bauer.)

ness, and the pollinia would be rendered absolutely useless. That insects often remove one
alone of the two pollinia at a time in many
kinds of Orchis is certain; it is even probable
that they generally remove only one at a
time, for the lower and older flowers almost
always have both pollinia removed, and the
younger flowers close beneath the buds, which
will have been seldomer visited, have frequently only one pollinium removed. In a
spike of Orchis maculata I found as many as
ten flowers, chiefly the upper ones, which had
only one pollinium removed; the other pollinium being in place, with the lip of the rostellum well closed up, and all the mechanism
perfect for its subsequent removal by some
insect.

The description now given of the action
of the organs in Orchis mascula applies to
O. morio, fusca, maculata, and latifolia, and to
Aceras anthropomorpha.* These species pre-

---

* The separation of this genus is evidently artificial. It is a
true Orchis, but with a very short nectary. Dr. Weddell has
described ('Annales des Sci. Nat.,' 3° ser. 'Bot.,' tom. xviii. p. 6)
the occurrence of numerous hybrids, naturally produced, between
this Aceras and Orchis galeata.

sent slight and apparently co-ordinated differences in the length of the caudicle, in the direction of the nectary, in the shape and position of the stigma, but they are not worth detailing. In all, the pollinia undergo, after removal from the anther-cells, the curious movement of depression, which is so necessary to place them in a right position on the insect's head to strike the stigmatic surface of another flower. In Aceras the caudicle is unusually short; the nectary consists of two minute rounded depressions; the stigma is transversely elongated; the two viscid discs lie so close together within the rostellum that they affect each other's outline; this is worth notice, as a step towards the two becoming absolutely confluent, as in O. pyramidalis. Nevertheless, in Aceras a single pollinium is sometimes removed by insects, though more rarely than with the other species.

We now come to Orchis pyramidalis, one of the most highly organised species which I have examined, and which is ranked by several botanists in a distinct genus. The relative position of the parts (Fig. III.) is here consi-

derably different from what it is in O. mascula
and its allies.   There are two quite distinct
rounded stigmatic surfaces (s, s, A) placed on
each side of the pouch-formed rostellum. This
latter organ, instead of standing some height
above the nectary, is brought down (see side
view B) so as to overhang and partially to
close its orifice.   The ante-chamber to the
nectary formed by the union of the edges of
the labellum to the column, which is large
in O. mascula and its allies, is here small.
The pouch-formed rostellum is hollowed out
on the under side in the middle : it is filled
with fluid.   The viscid disc is single, of the
shape of a saddle (Figs. C and E), carrying
on its nearly flat top or seat the two caudicles
of the pollinia; of which the two truncated
ends firmly adhere to its upper surface.   Before
the membrane of the rostellum ruptures, it can
be clearly seen that the saddle-formed disc
forms part of the continuous surface of the
rostellum.   The disc is partially hidden and
kept damp (which is of great importance) by
the largely over-folded basal membranes of
the two anther-cells.   The upper membrane
of the disc consists of several layers of minute

Fig. III.

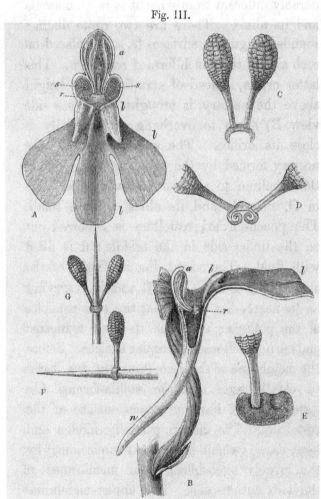

ORCHIS PYRAMIDALIS.

cells, and is therefore rather thick; it is lined beneath with a layer of highly adhesive matter, which is formed within the rostellum. The single saddle-formed disc strictly answers to the two minute, oval, separate discs of membrane to which the two caudicles of O mascula and its allies are attached: two separate discs have here become completely confluent.

When the flower opens and the rostellum has become symmetrically ruptured, either from a touch or spontaneously (I know not

---

Description of Fig. III.

a. anther.	*l.* labellum.
s. s. stigma.	*l'.* guiding plate on the labellum.
r. rostellum.	*n.* nectary.

A. Front view, with all sepals and petals removed, except the labellum.

B. Side view, with all sepals and petals removed, with the labellum longitudinally bisected, and with the near side of the upper part of the nectary cut away.

C. The two pollinia attached to the saddle-shaped viscid disc.

D. The disc after the first act of contraction, with no object seized.

E. The disc seen from above, and flattened by force, with one pollinium removed; showing the depression, by which the second act of contraction is effected.

F. The pollinium removed by the insertion of a needle into the nectary, after it has clasped the needle by the first act of contraction.

G. The same pollinium after the second act of contraction and depression.

which), the slightest touch depresses the lip,
that is, the lower and bilobed portion of the
exterior membrane of the rostellum, which projects into the mouth of the nectary. When the
lip is depressed, the under and viscid surface
of the disc, still remaining in its proper place,
is uncovered, and is almost certain to adhere
to the touching object. Even a human hair,
when pushed into the nectary, is stiff enough
to depress the lip or pouch; and the viscid surface of the saddle adheres to it. If, however,
the lip be touched too slightly, it springs back
and re-covers the under side of the saddle.

The perfect adaptation of the parts is well
shown by cutting off the end of the nectary
and inserting a bristle at that end; consequently in a reversed direction to that in
which nature intended moths to insert their
probosces, and it will be found that the rostellum may easily be torn or penetrated, but
that the saddle is rarely or never caught.
When the saddle sticking to a bristle together
with its pollinia is removed, the under lip
instantly curls closely inwards, and leaves the
orifice of the nectary more open than it was
before; but whether this is of any real use to

moths which so frequently visit the flowers, and consequently to the plant, I will not pretend to decide.

Lastly, the labellum is furnished with two prominent ridges (*l*, Fig. A, B), sloping down to the middle and expanding outwards like the mouth of a decoy; these ridges perfectly serve to guide any flexible body, like a fine bristle or hair, into the minute and rounded orifice of the nectary, which, small as it is, is partly choked up by the rostellum. This contrivance of the guiding ridges may be compared to the little instrument sometimes used for guiding a thread into the fine eye of a needle.

Now let us see how these parts act. Let a moth insert its proboscis (and we shall presently see how frequently the flowers are visited by Lepidoptera) between the guiding ridges of the labellum, or insert a fine bristle, and it is surely conducted to the minute orifice of the nectary, and can hardly fail to depress the lip of the rostellum; this being effected, the bristle comes into contact with the now naked and sticky under surface of the suspended saddle-formed disc. When the bristle is

c

removed, the saddle with the attached pollinia is removed. Almost instantly, as soon
as the saddle is exposed to the air, a rapid
movement takes place, and the two flaps curl
inwards and embrace the bristle. When the
pollinia are pulled out by their caudicles, by a
pair of pincers, so that the saddle has nothing
to clasp, I observed that the tips curled inwards
so as to touch each other in nine seconds (see
Fig. D), and in nine more seconds the saddle
was converted by curling still more inwards
into an apparently solid ball. The probosces of
the many moths which I have examined, with
the pollinia of this Orchis attached to them,
were so thin that the tips of the saddle just met
on the under side. Hence a naturalist, who sent
me a moth with several saddles attached to its
proboscis, and who did not know of this movement, very naturally came to the extraordinary conclusion that the moth had cleverly
bored through the exact centres of the so-called
sticky glands of some Orchid.

Of course this rapid clasping movement
helps to fix the saddle with its pollinia upright
on the proboscis, which is very important; but
the viscid matter rapidly setting hard would

probably suffice for this end, and the real object gained is the divergence of the pollinia. The pollinia, being attached to the flat top or seat of the saddle, project at first straight up and are nearly parallel to each other; but as the flat top curls round the cylindrical and thin proboscis, or round a bristle, the pollinia necessarily diverge. As soon as the saddle has clasped the bristle and the pollinia have diverged, a second movement commences, which action, like the last, is exclusively due to the contraction of the saddle-shaped disc of membrane, as will be more fully described in the seventh chapter. This second movement is the same as that in O. mascula and its allies, and causes the divergent pollinia, which at first projected at right angles to the needle or bristle (see Fig. F), to sweep through nearly 90 degrees towards the tip of the needle (see Fig. G), so as to become depressed and finally to lie in the same plane with the needle. In three specimens this second movement was effected in from 30 to 34 seconds after the removal of the pollinia from the anther-cells, and therefore in about 15 seconds after the saddle had clasped the bristle.

The use of this double movement becomes evident if a bristle with pollinia attached to it, which have diverged and become depressed, be pushed between the guiding ridges of the labellum into the nectary of the same or another flower (compare Figs. A and G); for the two ends of the pollinia will be found to have acquired exactly such a position that the end of the one strikes against the stigma on the one side, and the end of the other at the same moment strikes against the stigma on the opposite side. These stigmas are so viscid that they rupture the elastic threads by which the packets of pollen are bound together; and some dark-green grains will be seen, even by the naked eye, remaining on the two white stigmatic surfaces. I have shown this little experiment to several persons, and all have expressed the liveliest admiration at the perfection of the contrivance by which this Orchid is fertilised.

As in no other plant, or indeed in hardly any animal, can adaptations of one part to another, and of the whole to other organised beings widely remote in the scale of nature, be named more perfect than those presented by this Orchis, it may be worth while briefly

to sum them up. As the flowers are visited
both by day and night-flying Lepidoptera, I
do not think that it is fanciful to believe
that the bright-purple tint (whether or not
specially developed for this purpose) attracts
the day-fliers, and the strong foxy odour the
night-fliers. The upper sepal and two upper
petals form a hood protecting the anther and
stigmatic surfaces from the weather. The
labellum is developed into a long nectary in
order to attract Lepidoptera, and we shall
presently give reasons for suspecting that the
nectar is purposely so lodged that it can be
sucked only slowly (very differently from in
most flowers of other families), in order to give
time for the curious chemical quality of the
viscid matter on the under side of the saddle
setting hard and dry. He who will insert a
fine and flexible bristle into the expanded
mouth of the sloping ridges on the labellum
will not doubt that they serve as guides; and
that they effectually prevent the bristle or proboscis from being inserted obliquely into the
nectary. This circumstance is of manifest importance, for, if the proboscis were inserted
obliquely, the saddle-formed disc would become

attached obliquely, and after the compounded movement of the pollinia they could not strike the two lateral stigmatic surfaces.

Then we have the rostellum partially closing the mouth of the nectary, like a trap placed in a run for game ; and the trap so complex and perfect, with its symmetrical lines of rupture forming the saddle-shaped disc above, and the lip of the pouch below; and, lastly, this lip so easily depressed that the proboscis of a moth could hardly fail to uncover the viscid disc and adhere to it. But if this did fail to occur, the elastic lip would rise again and re-cover and keep damp the viscid surface. We see the viscid matter within the rostellum attached to the saddle-shaped disc alone, and surrounded by fluid, so that the viscid matter does not set hard till the disc is withdrawn.   Then we have the upper surface of the saddle, with its attached caudicles, also kept damp within the bases of the anther-cells, until withdrawn, when the curious clasping movement instantly commences, causing the pollinia to diverge, followed by the movement of depression, which compounded movements together are exactly fitted to cause the ends of the two pollinia to strike the two stigmatic surfaces. These stigmatic surfaces are sticky enough not to tear off the whole pollinium from the proboscis of the moth, but by rupturing the elastic threads to secure a few packets of pollen, leaving plenty for other flowers.

But let it be observed that, although the moth probably takes a considerable time to suck the nectar of any one flower, yet the movement of depression in the pollinia does not commence (as I know by trial) until the pollinia are fairly withdrawn out of their cells; nor will the movement be completed, and the pollinia be fitted to strike the stigmatic surfaces, until about half a minute has elapsed, which will give ample time for the moth to fly to another plant, and thus effect a union between two distinct individuals. Lastly, we have the wonderful growth of the pollen-tubes and their penetration of the stigma, as well as the mysteries of germination, though these are common to all phanerogamic plants.

Orchis ustulata* resembles O. pyramidalis

* I am greatly indebted to Mr. G. Chichester Oxenden of Broome Park for fresh specimens of this Orchis, and for his

in some important respects, and differs in
others. The labellum is deeply channelled;
this channel, which replaces the guiding ridges
of O. pyramidalis, leads to the small triangular orifice of the short nectary. The upper
angle of the triangle is overhung by the rostellum, the pouch of which is rather pointed
below. Owing to this position of the rostellum,
close to the mouth of the nectary, the stigma
is necessarily double and lateral; but we here
have an interesting gradation, showing how
easily the single and slightly lobed medial
stigma of O. maculata would pass through
the bilobed stigma of O. mascula into that of
O. ustulata, and thence into the truly double
stigma of O. pyramidalis; for in O. ustulata,
directly under the rostellum, there is a narrow
rim, in direct continuity with the two lateral
stigmas, and which itself has the character of
a true stigma, as it is formed of utriculi, or
true stigmatic tissue, exactly like that of the
lateral stigmas. The viscid discs are somewhat elongated. The pollinia undergo the

---

never-tiring kindness in supplying me with living plants, with
numerous specimens, and information regarding many of the
rarer British Orchids.

usual movement of depression, and in acquiring this position the two diverge slightly, so as to be ready to strike the lateral stigmas.

The divergence seemed due to the manner or direction in which the membrane forming the top of the disc contracted obliquely; but I am not sure of this observation.

I have now described the structure, as seen in fresh specimens, of most of the British species of the genus Orchis. All these species absolutely require the aid of insects for their fertilisation. This is obvious from the fact that the pollinia are so closely embedded in their anther-cells, and the disc with its ball of viscid matter in the pouch-formed rostellum, that they cannot be shaken out by violence. We have also seen numerous contrivances by which the pollinia assume, after an interval of time, a position adapted to strike the stigmatic surface; and this indicates that the pollinia are habitually carried from one flower to another. But to prove that insects are necessary I covered up a plant of Orchis morio under a bell-glass, before any of its pollinia had been removed, leaving three adjoining plants

uncovered; I looked at the latter every morning, and daily found some of the pollinia
removed, till all were removed with the exception of the pollinia in one flower low down
on one spike, and with the exception of those
in one or two flowers at the apex of each
spike, which were never removed. I then
looked at the perfectly healthy plant under
the bell-glass, and it had, of course, all its
pollinia in their cells. I tried an analogous
experiment with specimens of O. mascula with
exactly the same result. It deserves notice
that the spikes which had been covered up,
when subsequently left uncovered, had not
their pollinia removed, and did not, of course,
set any seed, whereas the adjoining plants
produced plenty of seed; and from this fact
I infer that probably there is a proper season
for each kind of Orchis, and that insects
cease their visits after the proper season has
passed, and the regular secretion of nectar
has ceased.

I have been in the habit for twenty years
of watching Orchids, and have never seen
an insect visit a flower, excepting butterflies
twice sucking O. pyramidalis and Gymnadenia

conopsea. That bees sometimes visit Orchids *
I have evidence in a humble and hive bee
sent me by Professor Westwood, with pollinia
attached to them; and Mr. F. Bond informs
me that he has seen pollinia attached to other
species of bees; but I feel almost certain that
bees do not habitually visit the common British
species of Orchis. On the other hand, I have
met with several accounts in entomological
works of pollinia having been observed attached
to moths. Mr. F. Bond was so kind as to send
me a large number of moths in this condition,
with permission, at the risk of the destruction
of the specimens, to remove the pollinia; and
this is quite necessary, in order to ascertain to
what species the pollinia belong. Singularly
all the pollinia (with the exception of a few
from Orchids of the genus Habenaria, pre-

* M. Ménière (in Bull. Bot. Soc. de France, tom. i. 1854,
p. 370) says he saw, in Dr. Guépin's collection, bees collected at
Saumur with the pollinia of Orchids attached to their heads; and
he states that a person who kept bees near the Jardin de la
Faculté (at Toulouse?) complained that his bees returned from
the garden with their heads charged with yellow bodies, of which
they could not free themselves. This is good evidence how
firmly the pollinia become attached. There is nothing to show
whether the pollinia in these cases belonged to the genus Orchis
or to other genera of the family, some of which I know are
visited by bees.

sently to be mentioned) belonged to O. pyramidalis. I here give the list of twenty-three
species of Lepidoptera, with the pollinia of O.
pyramidalis attached to their probosces.

Polyommatus alexis.	Eubolia mensuraria (two speci-
Lycæna phlæas.	mens).
Arge galathea.	Hadena dentina.
Hesperia sylvanus	Heliothis marginata (two
„ linea.	specimens).
Syrichthus alveolus.	Xylophasia sublustris (two
Anthrocera filipendulæ.	specimens).
„ trifolii.*	Euclidia glyphica.
Lithosia complana.	Toxocampa pastinum.
Leucania lithargyria (two	Melanippe rivaria.
specimens).	Spilodes palealis.
Caradrina blanda.	„ cinctalis.
„ alsines.	Acontia luctuosa.
Agrotis cataleuca.	

A large majority of these moths and butterflies had two or three pairs of pollinia attached
to them, and invariably to the proboscis. The
Acontia had seven pair, and the Caradrina no
less than eleven pair! The probosces of these
two latter moths presented an extraordinary

* I am indebted to Mr. Parfitt for an examination of this moth,
which is mentioned in the 'Entomologist's Weekly Intelligencer,'
vol. ii p. 182, and vol. iii. p. 3, Oct. 3, 1857. The pollinia were
erroneously thought to belong to Ophrys apifera. The pollen had
changed from its natural green colour to yellow; on washing it,
however, and drying it, the green tint returned.

arborescent appearance (Fig. IV.) The saddle-
formed discs adhered to
the proboscis, one be-
fore the other, with
perfect symmetry (as
necessarily follows from
its insertion having
been guided by the
ridges on the label-
lum), each saddle bear-
ing its pair of pollinia.

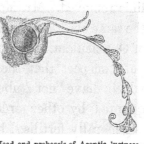

Fig. IV.

Head and proboscis of Acontia luctuosa
with seven pair of the pollinia of Orchis
pyramidalis attached to the proboscis.

The unfortunate Caradrina, with its proboscis
thus encumbered, could hardly have reached
the extremity of the nectary, and would soon
have been starved to death. These two moths
must have sucked many more than the seven
and eleven flowers, of which they bore the
trophies, for the earlier attached pollinia had
lost much of their pollen, showing that they
had touched many viscid stigmas.

This list shows, also, how many species of
Lepidoptera visit the same kind of Orchis.
The Hadena also frequents Habenaria. Probably all the Orchids provided with spur-like
nectaries are visited indifferently by many
kinds of moths. I have twice observed Gym-

nadenia conopsea, transplanted many miles from its native home, with nearly all its pollinia removed. Mr. Marshall of Ely* has made the same observation on transplanted specimens of O. maculata. I have not sufficient evidence, but I suspect that the Neotteæ and Malaxeæ, which have not tubular nectaries, are frequented by other orders of insects. Listera is generally fertilised by small Hymenoptera ; Spiranthes by humble-bees. Mr. Marshall found that fifteen plants of Ophrys muscifera, transplanted to Ely, had not one pollen-mass removed ; so it was during the first summer with Epipactis latifolia planted in my own garden ; during the following summer six flowers out of ten had their pollinia removed by some insect. These facts possibly indicate that certain Orchids require special insects for their fertilisation. On the other hand, Malaxis paludosa, placed in a bog about two miles from that in which it grew, had most of its pollinia immediately removed.

The list which follows serves to show that

---

* 'Gardener's Chronicle,' 1861, p. 73. Mr. Marshall's communication was in answer to some remarks of mine on the subject previously published in the ' Gardener's Chronicle,' 1860, p. 528.

in most cases moths perform the work of fertilisation effectually. But the list by no means gives a fair idea how effectually it is done; for I have often found nearly all the pollinia removed; but generally I kept an exact record in exceptional cases alone, as may be seen by the appended remarks. Moreover, in most cases, the pollinia which had not been removed were in the upper flowers beneath the buds, and many of these would probably have been subsequently removed. I have often found abundance of pollen on the stigmas of flowers which had not their own pollinia removed, showing that they had been visited by insects; in many other flowers the pollinia had been removed, but no pollen had, as yet, been left on their stigmas.

In the second lot of O. morio, given in the list, we see the injurious effects of the extraordinary cold and wet season of 1860 in the infrequency of the visits of insects, and, consequently, on the fertilisation of this Orchid. Very few seed-capsules were produced this year.

In O. pyramidalis I have examined spikes in which every single expanded flower had its

	Number of flowers with both or one pollinium removed. Flowers lately open excluded.	Number of flowers with only one pollinium removed. These flowers are included in the column to the left.	Number of flowers with neither pollinium removed.
Orchis morio. Three small plants. N. Kent	22	2	6
Thirty-eight plants. N. Kent. These plants were examined after nearly four weeks of extraordinarily cold and wet weather in 1860; and therefore under the most unfavourable circumstances	110	23	193
Orchis pyramidalis. Two plants. N. Kent and Devonshire	39	..	8
Six plants from two protected valleys. Devon	102	..	66
Six plants from a much exposed bank. Devon	57	..	166
Orchis maculata. One plant. Staffordshire. Of the twelve flowers which had not their pollinia removed, the greater number were young flowers under the buds	32	6	12
One plant. Surrey	21	5	7
Two plants. N. and S. Kent	28	17	50
Orchis latifolia. Nine plants from S. Kent, sent me by the Rev. B. S. Malden. The flowers were all mature	50	27	119
Orchis fusca. Two plants. S. Kent. Flowers quite mature, and even withered	8	5	54
Aceras anthropomorpha. Four plants. S. Kent	63	6	34

pollinia removed. The forty-nine lower flowers of a spike from Folkestone (sent me by Sir Charles Lyell) actually produced forty-eight fine seed-capsules; and of the sixty-nine lower flowers in three other spikes, seven alone had failed to produce capsules. These facts show conclusively how well moths had performed their office of marriage-priests.

In the list, the third lot of O. pyramidalis grew on a steep grassy bank, overhanging the sea near Torquay, and where there were no bushes or other shelter for moths; being surprised how few pollinia had been removed, though the spikes were old, and very many of the lower flowers had withered, I gathered, for comparison, six other spikes from two bushy and sheltered valleys, half a mile on each side of the exposed bank; these spikes were certainly younger, and would probably have had several more of their pollinia removed; but in their present condition we see how much more frequently they had been visited by moths, and consequently fertilised, than those growing on the much exposed bank. The Bee Ophrys and O. pyramidalis, in many parts of England, grow mingled together; and they did so here, but the Bee Ophrys, instead of being, as usual, the rarer species, was here much more abundant than O. pyramidalis; no one would readily have suspected that probably one chief reason of this difference was, that the exposed situation was unfavourable to moths, and therefore to the seeding of O. pyramidalis; whereas, as we shall hereafter see, the Bee Ophrys is independent of insects.

I counted many spikes of O. latifolia, because, being familiar with the usual state of the closely-allied O. maculata, I was surprised to observe in nine nearly withered spikes how few pollinia had been removed. In one instance, however, I found O. maculata even worse fertilised; for seven spikes, which had borne 315 flowers, produced only forty-nine seed-capsules—that is, each plant on an average produced only seven capsules: in this case the plants had grown in greater numbers close together, forming large beds, than I had ever before observed; and I imagined that there were too many plants for the moths to suck and fertilise. On some other plants, growing at no great distance, I found above thirty capsules on each spike.

Orchis fusca offers a more curious case of imperfect fertilisation. I examined ten fine spikes from two localities in South Kent, sent to me by Mr. Oxenden and Mr. Malden : most of the flowers on these spikes were partly withered, with the pollen mouldy even in the uppermost flowers; hence we may safely infer that no more pollinia would have been removed. I examined all the flowers only in two spikes, on account of the trouble from their withered condition, and the result may be seen in the list, namely, fifty-four flowers with both pollinia in place, and only eight with one or both removed. We see in this Orchid, and in O. latifolia, neither of which had been sufficiently visited by moths, that there are more flowers with one pollinium than with both removed. I casually examined many flowers in the other spikes of O. fusca, and the proportion of pollinia removed was evidently not greater than in the two given in the list. The ten spikes had borne 358 flowers, but, in accordance with the few pollinia removed, only eleven capsules had been formed : five of the ten spikes bore not a single capsule; two spikes had only one, and one bore as many as four capsules. As corroborating what I have previously said on pollen being often found on the stigmas of flowers which have their own pollinia in place, I may add that, of the eleven flowers which had produced capsules, five had both pollinia still within their now withered anther-cells.

From these facts the suspicion naturally arises that O. fusca is so rare a species in Britain from not being sufficiently attractive to our moths, and consequently not producing a sufficiency of seed. C. K. Sprengel * noticed, that in Germany O. militaris (ranked by Bentham as the same species with O. fusca) is likewise imperfectly fertilised, but more perfectly than our O. fusca ; for he found five old spikes bearing 138 flowers, which had set thirty-one capsules ; and he contrasts the state of these flowers with those of Gymnadenia conopsea, in which almost every flower produces a capsule.

An allied curious subject remains to be discussed. The existence of a well-developed spur-like nectary seems almost to imply the

* 'Das entdeckte Geheimniss,' etc. s. 404.

secretion of nectar. But Sprengel, a most
careful observer, thoroughly searched many
flowers of O. latifolia and of O. morio, and
could never find a drop of nectar; nor could
Krünitz* find nectar either in the nectary or
on the labellum of O. morio, fusca, militaris,
maculata, and latifolia. I have looked to all
the species hitherto mentioned in this work,
and could find no signs of nectar ; I examined,
for instance, eleven flowers of O. maculata,
taken from different plants growing in different districts, and taken from the most
favourable position on each spike, and could
not find under the microscope the smallest
bead of nectar. Sprengel calls these flowers
"Scheinsaftblumen," or sham-nectar-producers;
that is, he believes, for he well knew that the
visits of insects were indispensable for their
fertilisation, that these plants exist by an
organized system of deception. But when we
reflect on the incalculable number of plants
which have existed for enormous periods of
time, all absolutely requiring for each genera-

* Quoted by J. G. Kurr in his 'Untersuchungen über die
Bedeutung der Nektarien,' 1833, s. 28. See also 'Das entdeckte
Geheimniss,' s. 403.

tion insect - agency; when we think of the
special contrivances clearly showing that, after
an insect has visited one flower and has been
cheated, it must almost immediately go to
a second flower, in order that impregnation
may be effected (of which fact we have the
plainest evidence in the large number of pollinia attached to the probosces of those moths
which had visited O. pyramydalis), we cannot
believe in so gigantic an imposture. He who
believes in this doctrine must rank very low
the instinctive knowledge of many kinds of
moths.

To test the intellect of moths I tried the
following little experiment, which ought to
have been tried on a larger scale. I removed
a few already-opened flowers on a spike of
O. pyramidalis, and then cut off about half
the length of the nectaries of the six next
not-expanded flowers. When all the flowers
were nearly withered, I found that thirteen
of the fifteen upper flowers with perfect
nectaries had their pollinia removed, and two
alone had their pollinia still in their anther-
cells; of the six flowers with their nectaries
cut off, three had their pollinia removed, and

three were still in place; and this seems to indicate that moths do not go to work in a quite senseless manner.

Nature may be said to have tried, but not quite fairly, this same experiment; for Orchis pyramidalis, as shown by Mr. Bentham, * often produces monstrous flowers without a nectary, or with a short and imperfect one. Sir C. Lyell sent me several spikes from Folkestone with many flowers in this condition: I found six without a vestige of a nectary, and their pollinia had not been removed. In about a dozen other flowers, having either short nectaries, or with the labellum imperfect, with the guiding ridges either absent or developed in excess and rendered foliaceous, the pollinia in one alone had been removed, and the ovarium of another flower was swelling. Yet I found that the saddle-formed discs in the first six and in the dozen other flowers were perfect, and that they readily clasped a needle when inserted in the proper place. Moths had removed the pollinia, and had thoroughly well fertilised the perfect flowers on the same spikes; so that they must have

* 'Handbook of the British Flora,' 1858, p. 501.

neglected the monstrous flowers, or, if visiting them, the derangement in the complex mechanism had hindered the removement of the pollinia, and prevented their fertilisation.

From these several facts I still suspected that nectar must be secreted by our common Orchids, and I determined to examine O morio rigorously. As soon as many flowers were open, I began to examine them for twenty-three consecutive days : I looked at them after hot sunshine, after rain, and at all hours : I kept the spikes in water, and examined them at midnight, and early the next morning : I irritated the nectaries with a bristle, and exposed them to irritating vapours : I took flowers which had quite lately had their pollinia removed by insects, of which I had independent proof on one occasion by finding within the nectary grains of some foreign pollen ;* and I took other flowers which from their position on the spike would soon have had their pollinia removed; but the nectary was invariably quite dry.

* I may mention that on soaking and separating the laminæ of the proboscis of a moth, which had the pollinia of a Habenaria attached to its head, a surprising number of pollen-grains of some other plant were seen in the water.

I still thought that the secretion might perhaps take place at the earliest dawn, as I have found that the secretion of nectar in flowers of other orders ceases and commences in the most rapid manner. Consequently, as O pyramidalis is visited (as may be seen in the foregoing list) by butterflies and by several day-flying moths (such as Anthrocera and Acontia), I carefully examined its nectary, taking plants from several localities and the most likely flowers, as just explained; but the glittering points within the nectary were absolutely dry. Hence we may safely conclude that the nectaries of the above-named Orchids neither in this country nor in Germany ever contain nectar.

In examining the nectaries of O. morio and maculata, and especially of O. pyramidalis, I was surprised at the degree to which the inner and outer membranes forming the tube or spur were separated from each other,—also at the delicate nature of the inner membrane, which could be most easily penetrated,—and, lastly, at the quantity of fluid contained between these two membranes. So copious is this fluid, that, having at first merely cut off the

D

ends of the nectaries of O. pyramidalis, and
gently squeezing them on glass under the
microscope, such large drops of fluid exuded
from the cut ends that I concluded that the
nectaries certainly did contain nectar ; but
when I carefully made, without any pressure,
a slit along the upper surface, and looked into
the tube, I found that the inner surface was
quite dry.

I then examined the nectaries of Gymnadenia conopsea (a plant ranked by some
botanists as a true Orchis) and of Habenaria
bifolia, which are always one-third or two-
thirds full of nectar : the inner membrane presented the same structure in being covered
with papillæ, but there was a plain difference
in the inner and outer membranes being closely
united, instead of being, as with the above-
named species of Orchis, in some degree separated from each other and charged with fluid.
Hence I am led to suspect that moths penetrate the lax inner membrane of the nectaries
of these Orchids, and suck the copious fluid
between the two membranes. I am aware that
this is a bold hypothesis; for no case is recorded of nectar being contained between the

two membranes of a nectary,* or of Lepidoptera penetrating with their delicate probosces
even the laxest membrane.

We have seen how numerous and beautifully
adapted the contrivances are for the fertilisation of Orchids. We know that it is of the
highest importance that the pollinia, when
attached to the head or proboscis of an insect,
should not fall sideways or backwards. We
know that the ball of viscid matter at the
extremity of the pollinium rapidly becomes
more and more viscid, and sets hard in a few
minutes' time : therefore we can see that it
would be an advantage to the plant if the
moth were delayed in sucking the nectar, so
as to give time for the viscid disc to become
immoveably affixed. Assuredly moths would
be delayed if they had to bore through several
points of the inner membrane of the nectary,
and to suck the nectar from the intercellular
spaces. This explanation of the good thus
gained in some degree corroborates the hypo-

* The nearest approach to this supposed case, yet really distinct, is the secretion of nectar in several monocotyledonous plants
(as described by Ad. Brongniart in Bull. Soc. Bot. de France,
tom. i. 1854, p. 75) from between the two walls (feuillets) which
form the divisions of the ovarium. But the nectar in this case is
conducted to the outside by a channel ; and the secreting surface
is homologically an exterior surface.

thesis that the nectaries of the above-named
species of Orchis do not secrete nectar externally, but into internal cavities.

The following singular relation supports
this view more strongly. I have found nectar
within the nectaries of only five British species
of Ophreæ, namely, in Gymnadenia conopsea
and albida, in Habenaria bifolia and chlorantha, and in Peristylus (or Habenaria)
viridis. The first four of these species have
the viscid surface of the discs of their pollinia,
not enclosed within a pouch, but naked, which
by itself shows that the viscid matter has a
different chemical nature from that in the
species of true Orchis, and does not rapidly
set hard when exposed to the air. But to
make sure of this I removed the pollinia from
their anther-cells, so that the upper as well
as the under surfaces of the viscid discs were
freely exposed to the air; in Gymnadenia
conopsea the disc remained sticky for two
hours, and in Habenaria chlorantha for more
than twenty-four hours. In Peristylus viridis
the viscid disc is covered by a pouch-formed
membrane, but this is so minute that botanists
have overlooked it. I did not, when examining this species, see the importance of exactly

ascertaining how rapidly the viscid matter set hard; but I copy from my notes the words written at the time: "Disc remains sticky for some time when removed from its little pouch."

Now the bearing of these facts is clear: if, as is certainly the case, the viscid matter of the discs of these five latter species is so viscid as to serve at once for the firm attachment of the pollinia to insects, and does not quickly become more and more viscid and set hard, there could be no use in moths being delayed in sucking the nectar by having to bore through the inner membrane of the nectaries at several points; and in these five species, and in these alone, we find copious nectar ready stored for their use in the open tubular nectaries. If this relation, on the one hand, between the viscid matter requiring some little time to set hard, and the nectar being so lodged that moths are delayed in getting it; and, on the other hand, between the viscid matter being at first as viscid as ever it will become, and the nectar lying all ready for rapid suction, be accidental, it is a fortunate accident for the plant. If not accidental, and I cannot believe it to be accidental, what a singular case of adaptation!

CHAPTER II.

Ophreæ continued — Fly and Spider Ophrys — Bee Ophrys, apparently adapted for perpetual self-fertilisation, but with paradoxical contrivances for intercrossing — The Frog Orchis; fertilisation effected by nectar secreted from two parts of the labellum — Gymnadenia conopsea — Greater and Lesser Butterfly Orchis; their differences and means of fertilisation — Summary on the powers of movement in the pollinia.

WE now come to those genera of Ophreæ which chiefly differ from Orchis in having two separate pouch-formed rostellums,* instead of the two being confluent, as in Orchis. First for the genus Ophrys.

* It is not correct to speak of two rostellums, but the inaccuracy may be forgiven from its convenience. The rostellum strictly is a single organ, formed from the modification of the dorsal stigma and pistil; so that in Ophrys the two pouches and the intermediate space together form the true rostellum. Again, in Orchis I have spoken of the pouch-formed organ as the rostellum, but strictly the rostellum includes the little crest or fold of membrane projecting between the bases of the anther-cells. This folded crest (sometimes converted into a solid ridge) corresponds with the smooth ridge lying between the two pouches in Ophrys, and owes its protuberant and folded condition in Orchis to the two pouches having been brought together and be-become confluent. This modification will be more fully explained in the seventh chapter.

Fig. V.

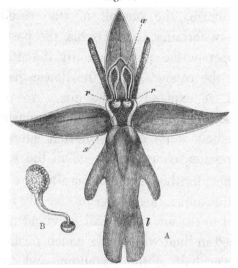

Ophrys muscifera, or Fly Ophrys.

a. anther.	*s.* stigma.
r r. rostellum.	*l.* labellum.

A. Flower viewed in front: the two upper petals are almost cylindrical and hairy: the two rostellums stand a little in advance of the bases of the anther-cells; but this is not shown from the foreshortening of the drawing.

B. One of the two pollinia removed from its anther-cell, and viewed laterally.

In *Ophrys muscifera*, or the Fly Ophrys, the chief peculiarity is that the caudicle of the pollinium (B) is doubly and almost rectangularly bent. The nearly circular piece of membrane, to the under side of which the ball of viscid matter is attached, is of considerable size, and plainly forms the summit of the rostellum, instead of forming, as in Orchis, the posterior and upper surface; consequently the attached end of the caudicle, after the flower has expanded, is exposed to the air. As might have been expected from this circumstance, the caudicle is not capable of that movement of depression, characteristic of all the species of Orchis; for this movement is always excited when the upper membrane of the disc is first exposed to the air. The ball of viscid matter is bathed in fluid within the pouch formed by the lower half of the rostellum, and this is necessary, as the viscid matter rapidly sets hard in the air. The pouch is not elastic, and does not spring up when the pollinium is removed. Such elasticity would have been of no use, as there is here a separate pouch for each viscid disc; whereas in Orchis, after one pollinium has been removed, the other has to be kept covered up and ready for action. Hence it would appear as if nature were so economical as to save even superfluous elasticity.

The pollinia cannot, as I have often tried,

be jarred out of the anther-cells by violence.
That insects of some kind visit these flowers,
though not frequently, and remove the pollinia, is certain, as we shall immediately see.
Twice I have found abundant pollen on the
stigmas of flowers, in which both their own
pollinia were still in their cells; and, no doubt,
had I oftener looked, I should have oftener
observed this fact. The elongated labellum
affords a good standing-place for insects : at
its base, just beneath the stigma, there is a
rather deep depression, representing the nectary in Orchis; but I could never see a trace
of nectar; nor have I observed any insects,
often as I have watched these inconspicuous
and scentless flowers, even approach them.
On each side of the base of the labellum there
is a shining knob, with an almost metallic
lustre, appearing like two drops of fluid ; and
if I could in any case believe in Sprengel's
sham-nectaries, I should believe it in this
instance. What induces insects to visit these
flowers I can at present only conjecture. The
two pointed pouches, covering the viscid discs,
stand not far apart, and project over the
stigma : any object gently pushed right against

D 3

one of them (in Orchis the push should be directed rather downwards) depresses the pouch, touches and adheres to the viscid ball, and the pollinium is easily removed.

The structure of the flower leads me to believe that small insects (as we shall see in the case of Listera) crawl along the labellum to its base, and that in bending their heads downwards or upwards they strike against one of the pouches; they then fly to another flower with a pollinium attached to their heads, and there bending down to the base of the labellum, the pollinium, owing to its doubly bent caudicle, strikes the sticky stigmatic surface, and leaves pollen on it. Under the next species we shall see good reason to believe that the doubly bent caudicle of the Fly Ophrys serves instead of the usual movement of depression.

That insects do visit the flowers of the Fly Ophrys and remove the pollinia, though not effectually or sufficiently, the following cases show. During several years before 1858 I occasionally examined some flowers, and found that out of 102 only thirteen had one or both pollinia removed. Although at the time I

recorded in my notes that most of the flowers
were partly withered, I now think that I must
have included a good many young flowers,
which might perhaps have been subsequently
visited ; so I prefer trusting to the following
observations.

	Number of Flowers.	
	Both Pollinia or one removed by Insects.	Both Pollinia in their Cells.
In 1858, 17 plants, bearing 57 flowers, growing near each other	30	27
In 1858, 25 plants growing in another spot, and bearing 65 flowers	15	50
In 1860, 17 plants, bearing 61 flowers	28	33
In 1861, 4 plants bearing 24 flowers from S. Kent, all the previous plants having grown in N. Kent	15	9
Total	88	119

We here see that, out of 207 flowers examined, not half had been visited by insects.
Of the eighty-eight flowers visited, thirty-one
had only one pollinium removed. As the visits
of insects are indispensable to the fertilisation
of this Orchid, it is remarkable (as in the case
of Orchis fusca) that this Ophrys has not been
rendered more attractive to insects.   The num-

ber of seed-capsules produced is proportionably
even less than the number of flowers visited
by insects. The year 1861 was extraordinarily favourable to this species in this part
of Kent, and I never saw such numbers in
flower ; accordingly I marked eleven plants,
which bore forty-nine flowers, but these produced only seven capsules. Two of the plants
each bore two capsules, and three other plants
each bore one, so that no less than six plants
did not produce a single capsule! What are
we to conclude from these facts? Are the
conditions of life unfavourable to this species,
though it was so numerous in some places
this year as to deserve being called quite common? Could the plant nourish more seed;
and would it be of any advantage to it to
produce more seed? Why does it produce so
many flowers, if a larger number of seeds
would not be advantageous to it? Something
seems to be out of joint in the machinery of
its life. We shall presently see what a remarkable contrast another species of this same
genus, the Ophrys apifera or Bee Ophrys, presents in producing seed.

Ophrys aranifera, or the Spider Ophrys.—I

am indebted to Mr. Oxenden for a few spikes of
this rare species. The
caudicle (Fig. A) rises
straight up from the
viscid disc, and is then
curved or bent forwards
in the same manner, but
not in so great a degree,
as in the last species.
The point of attachment

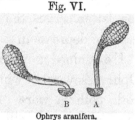

Fig. VI.

B   A

Ophrys aranifera.

A. Pollinium before the act of
depression.
B. Pollinium after the act of
depression.

of the caudicle to the membrane of the disc is
hidden within the bases of the anther-cells, and
is thus kept damp; consequently, as soon as
the pollinia are exposed to the air, the usual
movement of depression takes place, and the
pollinia sweep through an angle of about
ninety degrees. By this movement the pollinia (supposing them to have become attached
to an insect's head) assume a position exactly
adapted to strike the stigmatic surface, which
is situated, relatively to the pouch-formed rostellums, rather lower down in the flower than
in the Fly Ophrys. If we compare the woodcut of the pollinium of the present species, or
Spider Ophrys, after the movement, with that
of the Fly Ophrys, which is incapable of

movement, it is impossible to doubt that the permanent rectangular bend near the disc of the latter serves the same end as the movement of depression.

I examined fourteen flowers of the Spider Ophrys, several of which were partly withered; both pollinia were removed in none, and in three alone one pollinium had been removed. Hence this species, like the Fly Ophrys, apparently is not much visited by insects.

The anther-cells are remarkably open, so that, in travelling in a box, two pair of pollinia had fallen out, and were sticking by their viscid discs to the flower. Here we have, as throughout nature, evidence of gradation; for though the wide opening of the anther-cells is of no use to this species, it is of the highest importance, as we shall immediately see, to the following species, namely, the Bee Ophrys. So, again, the flexure of the upper end of the caudicle of the pollinium towards the labellum, though of service to the Spider and Fly Ophrys, in order that the pollinium, when removed by insects and carried to another flower, should strike the stigma, is exaggerated in the following species, and serves

for the very different purpose of self-fertilisation.

Ophrys apifera.—In the Bee Ophrys we
meet with widely different means of fertilisation as compared with the other species of the
genus, and, indeed, as far as I know, with all
other Orchids. The two pouch-formed rostellums, the viscid discs, and the position of the

Fig. VII.

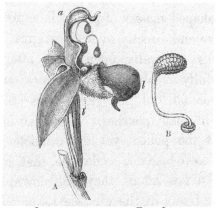

OPHRYS APIFERA, OR BEE OPHRYS.

a. anther.   l l. labellum.

A. Side view of flower, with the upper sepal and the two upper
    petals removed.   One pollinium, with its disc still in its
    pouch, is represented as just falling out of the anther-cell;
    and the other has fallen almost to its full extent, opposite
    to the hidden stigmatic surface.
B. Pollinium in the position in which it lies embedded.

stigma, are nearly the same as in other species
of Ophrys ; but to my surprise, I have observed
that the distance of the two pouches from each
other, and the shape of the mass of pollen-
grains, are variable. The caudicles of the
pollinia are remarkably long, thin, and flexible,
instead of being, as in all the other Ophreæ,
rigid enough to stand upright. They are
necessarily curved forward at their upper ends,
owing to the shape of the anther-cells ; and
the pear-shaped masses of pollen lie embedded
high above and directly over the stigma. The
anther-cells naturally open soon after the
flower is fully expanded, and the thick ends of
the pollinia fall out, the viscid discs still remaining in their pouches. Slight as is the
weight of the pollen, yet the caudicle is so
thin; and soon becomes so flexible, that, in the
course of a few hours, they sink down, until
they hang freely in the air (see lower pollen-
mass in Fig. A) exactly opposite to and in
front of the stigmatic surface. When in this
position a breath of air, acting on the expanded
petals, sets the flexible and elastic caudicles
vibrating, and they almost immediately strike
the viscid stigma, and, being there secured,

impregnation is effected. To make sure that
no other aid was requisite, though the experiment was superfluous, I covered up a plant
under a net, so that some wind, but no insects,
could pass in, and in a few days the pollinia
had become attached to the stigmas ; but the
pollinia of a spike kept in water in a still room,
remained free, suspended in front of the stigma.

Robert Brown * first observed that the
structure of the Bee Ophrys is adapted for self-
fertilisation. When we consider the unusual
and perfectly-adapted length, as well as the
remarkable thinness, of the caudicles of the
pollinia ; when we see that the anther-cells
naturally open, and that the masses of pollen,
from their weight, slowly fall down to the
exact level of the stigmatic surface, and are
there made to vibrate to and fro by the
slightest breath of wind till the stigma is
struck ; it is impossible to doubt that these
points of structure and function, which occur
in no other British Orchid, are specially
adapted for self-fertilisation.

* ' Transact. Linn. Soc.' vol. xvi. p. 740. Brown erroneously
believed that this peculiarity was common to the genus. As far
as the British species are concerned, it applies to this one alone
of the four species.

The result is what might have been anticipated. I have often noticed that the spikes of
the Bee Ophrys apparently produced as many
seed-capsules as flowers ; and near Torquay I
carefully examined many dozen plants, some
time after the flowering season ; and on all I
found from one to four, and occasionally five,
fine capsules, that is, as many capsules as there
had been flowers ; in extremely few cases
(excepting a few deformed flowers, generally
on the summit of the spike) could a flower be
found which had not produced a capsule. Let
it be observed what a contrast this case presents with that of the Fly Ophrys, which
requires insect agency, and which from forty-
nine flowers produced only seven capsules !

From what I had seen of other British
Orchids, I was so much surprised at the self-
fertilisation of this species, that during many
years I have looked at the state of the pollen-
masses in hundreds of flowers, and I have
never seen, in a single instance, reason to believe that pollen had been brought from one
flower to another. Excepting in a few monstrous flowers, I have never seen an instance
of the pollinia failing to reach their own stig-

mas. In a very few cases I have found one pollinium removed, but in some of these cases the marks of slime led me to suppose that slugs had devoured them. For instance, in 1860, I examined in North Kent twelve spikes bearing thirty-nine flowers, and three of these had one pollinium removed, all the other pollinia being glued to their own stigmas. In another lot, from another locality, however, I found the unparalleled case of two flowers with both pollinia removed, and two others with one removed. I have examined some flowers from South Kent with the same result. Near Torquay I examined twelve spikes bearing thirty-eight flowers, and in these one single pollinium alone had been removed. We must not forget that blows from animals or storms of wind might occasionally cause the loss of a pollinium.

In the Isle of Wight Mr. A. G. More was so kind as to examine carefully a large number of flowers. He observed that in plants growing singly both pollinia were invariably present. But on taking home several plants, from a large number growing in two places, and *selecting plants* which seemed to have had

some pollinia removed, he examined 136
flowers : of these ten had lost both pollinia,
and fourteen had lost one; so here there
seems at first evidence of the pollinia having
been removed by their adhesion to insects;
but then Mr. More found no less than eleven
pollinia (not included in the above cases of
removal) with their caudicles cut or gnawed
through, but with their viscid discs still in
their pouches, and this proves that some other
animals, not insects, probably slugs, had been
at work. Three of the flowers were much
gnawed. Two pollinia, which had apparently
been thrown out by strong wind, were sticking
to the sepals, and three pollinia were found
loose in his collecting box, so that it is very
doubtful whether many, or indeed any, of the
pollinia had been removed by adhesion to
insects. I will only add that I have never
seen an insect visit these flowers.* Robert
Brown imagined that the flowers resembled
bees in order to deter insects from visiting

---

* Mr. Gerard E. Smith, in his ' Catalogue of Plants of S. Kent,'
1829, p. 25, says : " Mr. Price has frequently witnessed attacks
made upon the Bee Orchis by a bee, similar to those of the
troublesome Apis muscorum." What this sentence means I cannot conjecture.

them; I cannot think this probable. The equal or greater resemblance of the Fly Ophrys to an insect does not deter the visits of some unknown insect, which, in that species, are indispensable for the act of fertilisation.

Whether we look to the structure of the several parts of the flower as far as hitherto described, or to the actual state of the pollinia in numerous plants taken during different seasons from different localities, or to the number of seed-capsules produced, the evidence seems conclusive that we here have a plant which is self-fertilised for perpetuity. But now let us look to the other side of the case. When an object is pushed (as in the case of the Fly Ophrys) right against one of the pouches of the rostellum, the lip is depressed, and the large extremely viscid disc adheres firmly to the object, and the pollinium is removed. Even after the pollinia have naturally fallen out of their cells and are glued to the stigma, their removal can sometimes be thus effected. As soon as the disc is drawn out of the pouch the movement of depression commences by which the pollinium would be brought in front of an insect's head ready to strike the stigma. When a pollen-mass is placed on the stigma and then withdrawn, the elastic threads by which the packets are tied together, break, and leave several packets on the viscid surface. In all other Orchids the meaning of these several contrivances—namely, the downward movement of the lip of the rostellum when gently pushed —the viscidity of the disc—the act of depression of the caudicle after the disc has been removed—the rupturing of the elastic threads by the viscidity of the stigma, so that pollen may be left on several stigmas—is unmistakably clear. Are we to believe that these contrivances in the Bee Ophrys are absolutely purposeless, as would certainly be the case if this species is perpetually self-fertilised? If the discs had been small or only viscid in a slight degree, if the other related contrivances had been imperfect in any degree, we might have concluded that they had begun to abort; that Nature, if I may use the expression, seeing that the Fly and Spider Ophrys were imperfectly fertilised and produced few seed-capsules, had changed her plan and effected complete and perpetual self-fertilisation, in order that more seeds might be produced.

The case is perplexing in an unparalleled degree, for in the same flower we apparently have elaborate contrivances for directly opposed objects.

We have already seen many curious structures and movements, as in Orchis pyramidalis, which evidently lead to the fertilisation of one flower by the pollen of another flower, and we shall meet with numerous other and very different contrivances for the same object throughout the whole great Orchidean Family. Hence it is impossible to doubt that some great good is derived from the union of two distinct flowers, often borne on distinct plants ; but the good in the case of the Fly and Spider Ophrys is gained at the expense of much lessened fertility. In the Bee Ophrys great fertility is gained at the expense of apparently perpetual self-fertilisation ; but the contrivances are still present which are assuredly adapted to give an occasional cross with another individual ; and the safest conclusion, as it seems to me, is, that under certain unknown circumstances, and perhaps at very long intervals of time, one individual of the Bee Ophrys is crossed by another. Thus the generative functions of this plant would be brought into harmony with those of other Orchidaceæ, and, indeed, with those of all other plants, as far as I have been able to make out their structure.

Ophrys arachnites.—This form is considered by some high botanical authorities as a mere variety of the varying Bee Ophrys. Mr. Oxenden sent me two spikes bearing seven flowers. The anther-cells do not stand so high above the stigma, and do not overhang it so much, as in the Bee Ophrys. The mass of pollen-grains is generally more elongated. The upper part of the caudicle is curved for-

Fig. VIII.

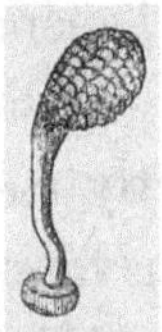

Pollinium of Ophrys arachnites.

ward, and the lower part under-goes the movement of depression, as in the Spider and Bee Ophrys. The caudicle in length compared with that of the Bee Ophrys is only as two to three, or even as two to four; though thus relatively shorter, it is as thick and broad as that of the Bee Ophrys: it is much more rigid, so that, when the upper end of the pollinium is forced out of its anther-cell, the sticky disc remaining in the pouch, it can only with difficulty be bent down to the stigma. We here see no adaptation for self-fertilisation.

The seven flowers sent me had evidently long been fully expanded, and the spikes, having travelled by railway, must have been well shaken, yet in six of the flowers both pollinia remained in their anther-cells; in the seventh, both pollinia adhered to the stigma with their discs still in their pouches; but this flower was much withered, and might have been crushed. Of the six flowers, three were so old that the pollen was mouldy and the petals discoloured; yet, as just stated, the pollinia were still in their cells. After having examined so many hundred flowers of the Bee Ophrys, I have never seen such a case. Considering this important functional difference in O. apifera and arachnites, and the lesser differences in the structure of their pollinia, which are likewise of functional signification, and the slight differences in their flowers, it seems to me that, until these forms can be shown to be connected by intermediate varieties, we must rank O. arachnites as a good species, more closely allied to O. aranifera in its manner of fertilisation than to O. apifera.

E

Herminium monorchis.—The Musk Orchis is generally spoken of as having naked glands or discs, but this is not strictly correct. The disc is of unparalleled size, nearly equalling the mass of pollen-grains : it is subtriangular in shape, unsymmetrical, somewhat resembling a helmet, but with one side protuberant : it is formed of hard membrane ; the hollow base alone is viscid, and this rests on and is covered by a narrow strip of membrane, which is easily pushed away, and answers to the pouch in Orchis. The whole upper part of the helmet answers to the minute oval bit of membrane to which the caudicle is attached in Orchis, and which is of larger size and convex in the Fly Ophrys. When the lower part of the helmet is moved by any pointed object, the point so readily slips into the hollow base, and is there so firmly held by the viscid matter, that it appears adapted to stick to some prominent part of an insect's head. The caudicle is short and very elastic ; it is attached not to the apex of the helmet, but to the hinder end ; if it had been attached to the apex, the point of attachment would have been freely exposed to the air, and would not have

contracted and caused the movement of depression in the pollinia when removed from
their anther-cells. This movement is well
marked, and is necessary to bring the end of
the pollen-mass into a proper position to
strike the stigma. The two viscid discs stand
wide apart. There are two transverse stigmatic surfaces, meeting by their points in the
middle; but the broad part of each lies directly
under the disc.

The labellum is upturned, making the flower
almost tubular. As far as I could ascertain,
an insect, in crawling out of, or into, the flower,
would be apt to strike against the upper
and extraordinarily protuberant ends of the
helmet-like discs, and so displace the inferior
viscid surfaces, which would adhere to its
head or body. There is so deep a hollow at
the base of the labellum as almost to deserve
to be called a nectary; but I could not perceive
any nectar. The flowers are very small and
inconspicuous, but emit a strong musky smell,
especially at night. They seem highly attractive to insects; in a spike with only seven
flowers recently open, four had both pollinia,
and one had a single pollinium removed.

Peristylus (or *Habenaria*) *viridis.* — The
Frog Orchis has also been described as having
its viscid discs naked, which is incorrect. The

Fig. IX.

Peristylus viridis, or Frog Orchis.
Front view of flower.

a. anther.
s. stigma.
n. orifice of central nectary.
n'. lateral nectaries.
l. labellum.

two small pouches stand
wide apart. The ball of
viscid matter is oval, and
does not soon set hard;
its surface is protected
by a minute pouch. The
upper membranous sur-
face of the disc is large,
and as in the Fly Ophrys
(O. muscifera) the point
of junction with the cau-
dicle is freely exposed
to the air, and does not
cause the pollinium to undergo the often described
movement of depression. But the caudicles
are not doubly bent as in the Fly Ophrys.
The stigmatic surface is small and medial;
and though the anther-cases are somewhat
inclined backwards and converge a little at
their upper ends, thus affecting the position
of the pollinia when attached to any object,
yet it is at first difficult to understand how the

pollinia, when removed by insects, can strike the stigma.

The explanation is rather curious. The elongated labellum forms a rather deep hollow in front of the stigma, and in this hollow, but some way in advance of the stigma, a minute slit-like orifice (*n*) leads into the short bilobed nectary. Hence an insect, in order to suck the nectar with which the nectary is filled, would have to bend down its head in front of the stigma. The labellum has a medial ridge, which would probably induce the insect to alight on either side ; but, apparently to make sure of this, besides the true nectary, there are two spots (*n'*) on each side at the base of the labellum, bordered by its prominent edges, directly under the two pouches, which secrete drops of nectar. Now let us suppose an insect to alight, probably on one side of the labellum, and first lick up the drop of nectar on either side; from the position of the pouches exactly over these drops, it would almost certainly get the pollinium of that side attached to its head; if it were then to go to the mouth of the true nectary, it would assuredly strike the pollinium against the stigma. So that we see this unique

case of nectar being secreted from the basal
edges of the labellum as well as within the
short medial nectary, replaces the power of
movement in the pollinia, so general with
other Orchids, and replaces the doubly bent
caudicles of the Fly Ophrys.

As I have described the case, the flower
would receive its own pollen; but if the insect
first exhausted the richer source of nectar
within the nectary, and afterwards licked up
the lateral drops, it would not till then get
the pollinia attached to its head, and, flying
to another flower, a union would be effected
between two distinct flowers or two distinct
plants. If, indeed, the insect were to suck the
lateral drops first, from what Sprengel has
observed in the case of Listera,* the insect
would perhaps be disturbed by the attachment of the pollinium, and would not go
on sucking immediately, but would fly to
another flower, and thus a union between
distinct individuals would ensue.

I am indebted to the Rev. B. S. Malden
of Canterbury for two spikes of the Frog
Orchis. Several of the flowers had one pol-

---

* 'Das endeckte Geheimniss der Natur,' s. 407.

linium removed, and one flower had both
removed.

We now come to two genera, namely,
Gymnadenia and Habenaria, including four
British species, which really have uncovered
viscid discs. The viscid matter, as before
remarked, is of a somewhat different nature
from that in Orchis, and does not rapidly set
hard. Their nectaries are stored with nectar.
With respect to the uncovered condition of the
discs, the last species, or Peristylus viridis, is
in an almost intermediate condition. The four
following forms compose a much broken series.
In Gymnadenia conopsea the viscid discs are
narrow and much elongated, and lie close
together; in G. albida they are less elongated,
but still approximate; in Habenaria bifolia
they are oval and far apart; and, lastly, in
Hab. chlorantha they are circular and much
farther apart.

Gymnadenia conopsea.—In general appearance this plant resembles pretty closely some
species of true Orchis. The pollinia differ
in having naked, narrow, strap-shaped discs,
which are nearly as long as the caudicles

(Fig. X.).   When the pollinia are exposed to

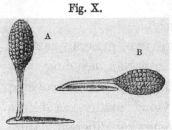

Fig. X.

Gymnadenia conopsea.
A. Pollinium, before the act of depression.
B.   „   after the act of depression, but
     before it has closely clasped the disc.

the air the caudicle is depressed in from 30 to 60 seconds; and as its anterior surface is slightly hollowed out, it closely clasps the upper membranous surface of the disc.

The mechanism of this movement will be described in the last chapter. The elastic threads by which the packets of pollen are bound together are unusually weak, as is likewise the case with the two following species of Habenaria : this was well shown by the state of specimens which had been kept in spirits of wine. This weakness apparently stands in relation to the viscid matter of the discs not setting hard and dry as in Orchis ; so that a moth with a pollinium attached to its proboscis might be enabled to visit several flowers and not have the whole pollinium dragged off by the first stigma which was struck. The two strap-shaped discs lie close together, and form the roof of the mouth of the nectary. They are not enclosed, as in Orchis, by a lower lip or pouch, so that the structure of the rostellum is simpler. When we come to treat of the homologies of the rostellum we shall see that this difference is due to a small change, namely, to the lower and exterior cells of the rostellum resolving themselves into viscid matter; whereas in Orchis the exterior surface retains its early cellular or membranous condition.

As the two viscid discs form the roof of the mouth of the nectary, and are thus brought down near to the labellum, the two stigmas, instead of being confluent and standing beneath the rostellum, are necessarily lateral and separate. They form two protuberant, almost horn-shaped, processes on each side of the mouth of the nectary. That their surfaces are really stigmatic I ascertained by finding them deeply penetrated by a multitude of pollen-tubes. As in the case of O. pyramidalis, it is a pretty little experiment to push a fine bristle into the narrow mouth of the nectary, and to observe how certainly the narrow elongated viscid discs, forming the roof, stick to the bristle. When the bristle is withdrawn, the pollinia are withdrawn, adhering to its upper side, and

E 3

slightly divergent owing apparently to their original position in the anther-cells. They then quickly depress themselves till they lie in the same plane with the bristle; and if the bristle, held in the same relative position, be now inserted into the nectary of another flower, the two ends of the pollinia accurately strike the two stigmatic surfaces lying close on each side of the mouth of the nectary. I am, however, not quite sure that I understand the cause of the divergence of the pollinia, for I find that moths often remove one pollinium alone; and this fact leads me to suspect that they insert their probosces obliquely into the nectary.

The flowers smell sweet, and the abundant nectar always contained in their nectaries seems highly attractive to Lepidoptera, for the pollinia are soon and effectually removed. For instance, in a spike with forty-five open flowers, forty-one had their pollinia removed, or had pollen left on their stigmas: in another spike with fifty-four flowers, thirty-seven had both pollinia, and fifteen had one pollinium, removed; so that only two flowers in the whole spike had neither pollinium removed.

Gymnadenia albida.—The structure of this

flower resembles in most points that of the last species ; but, owing to the upturning of the labellum, it is rendered almost tubular. The naked glands are minute, but elongated and approximate. The stigmatic surfaces are partly lateral and divergent. The nectary is short, and full of nectar. Minute as the flowers are, they seem highly attractive to insects : of the eighteen lower flowers in one spike, ten had both pollinia, and seven had one, removed ; in some other and older spikes all the pollinia had been removed, except from two or three flowers at the summit.

Habenaria chlorantha.—The pollinia of the Large Butterfly Orchis differ considerably from those of the species hitherto mentioned. The two anther-cells are separated from each other by a wide space of connective membrane ; the pollinia slope backwards (Fig. XI.), and the viscid discs are brought out in advance of the stigmatic surface, and front each other. In relation to this position of the discs, the caudicles and pollen-masses are much elongated. The viscid disc is circular, and, in the early bud, consists of a mass of cells, of which the exterior layers (answering to the lip or pouch

Fig. XI.

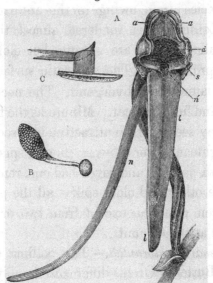

HABENARIA CHLORANTHA, OR LARGE BUTTERFLY ORCHIS.

a a. anther.	*n.* nectary.
d. disc.	*n'.* orifice of nectary.
s. stigma.	*l.* labellum.

A. Flower viewed in front, with all the sepals and petals removed except the labellum with its nectary.

B. A pollinium. (This has hardly a sufficiently elongated appearance.) The drum-like pedicel is hidden behind the disc.

C. Diagram, giving a section through the viscid disc, which is formed of an upper membrane with a layer of viscid matter beneath, and through the drum-like pedicel, and through the lower end of the caudicle.

in Orchis) resolve themselves into adhesive matter. This matter has the property of remaining adhesive for at least twenty-four hours after the pollinium has been removed from its cell. The disc, externally covered with a thick layer of adhesive matter (see Sect. C, which stands so that the layer of viscid matter is below), is produced on its opposite and embedded side into a short drum-like pedicel. This pedicel is continuous with the membranous portion of the disc, and is formed of the same tissue. At the embedded end of the pedicel, the caudicle of the pollinium is attached in a transverse direction, and its extremity is prolonged, as a bent rudimentary tail, just beyond the drum. The caudicle is thus united to the viscid disc in a very different manner, and in a plane at right angles, to what occurs in other British Orchids. In the short drum-like pedicel, we see a small development of the long pedicel of the rostellum, which in many exotic Vandeæ is so conspicuous, and which connects the viscid disc with the true caudicle of the pollinium.

The drum-like pedicel is of the highest importance, not only by rendering the viscid

disc more prominent and more likely to stick
to the face of an insect whilst inserting its
proboscis into the nectary beneath the stigma,
but on account of its power of contraction.
The pollinia lie inclined backwards in their
cells (see Fig. A), above and some way on
each side of the stigmatic surface; if attached
in this position to the head of an insect, the
insect might visit any number of flowers, and
no pollen could be left on the stigma. But
observe what takes place : in a few seconds
after the inner end of the drum-like pedicel is
removed from its imbedded position and exposed to the air, one side of the drum contracts, and this contraction draws the thick
end of the pollinium inwards, so that the caudicle and the viscid surface of the disc are no
longer parallel, as they were at first, and as
they are represented in the section C. At the
same time the drum rotates through nearly a
quarter of a circle, and this moves the caudicle
downwards, like the hand of a clock, depressing
the thick end of the pollinium or mass of pollen-
grains. After this double movement, the right-
hand disc, for instance, being supposed to be
affixed to the right side of an insect's face,

when the insect, after a short interval of time,
visits another flower, the pollen-bearing end
of the pollinium will have moved downwards
and inwards, and will now infallibly strike the
viscid surface of the stigma, situated in the
middle beneath and between the two anther-
cells.

The little rudimentary tail of the caudicle
projecting beyond the drum-like pedicel is an
interesting point to those who believe in the
modification of species; for it shows us that
the disc has moved a little inwards, and that
primordially the two discs stood even still
further in advance of the stigma than they do
at present.   It shows us that the parent-form
approached a little more closely in structure to
that extraordinary Orchid the Bonatia speciosa
of the Cape of Good Hope.

The remarkable length of the nectary, containing much nectar, the white colour of the
conspicuous flower, and the strong sweet odour
emitted at night, all show that this plant depends for its fertilisation on the larger nocturnal
Lepidoptera.   I have often found spikes with
almost all the pollinia removed.   From the
lateral position and distance of the two viscid

discs from each other, the same moth would generally remove only one pollinium at a time; and in a spike which had not as yet been much visited, three flowers had both pollinia, and eight flowers had only one pollinium, removed. From the position of the discs it might have been anticipated that they would adhere to the side of the head or face of moths; and Mr. F. Bond sent me a specimen of Hadena dentina with one eye covered and blinded by a disc, and a specimen of Plusia v. aureum with a disc attached to the edge of the eye. Although the discs are so extremely adhesive that almost all the pollinia of a bunch of flowers, when carried in the hand, will be removed by the shaken petals and sepals touching the discs, yet it is certain that moths, perhaps the smaller species, often visit these flowers without removing the pollinia; for on carefully examining the discs of a large number of pollinia still in their cells I found minute Lepidopterous scales glued to them.

Habenaria bifolia, or *Lesser Butterfly Orchis.* —I am aware that this form and the last are considered by Mr. Bentham and some other botanists as mere varieties of each other; for

it is said that intermediate gradations occur in
the position of the viscid discs. But we shall
immediately see that the two forms differ in a
great number of characters, not to mention the
differences in general aspect and in the stations
inhabited, with which we are not here concerned. Should these two forms be hereafter
proved to graduate at the present day into
each other, it would be a remarkable case of
variation; and I, for one, should be as much
pleased as surprised at the fact, for these two
forms certainly differ from each other more
than do most species of the genus Orchis.

The viscid discs of the Lesser Butterfly
Orchis are oval; they face each other, and
stand far closer together than in the last
species; so much so, that in the bud, when
their surfaces are cellular, they almost touch.
They are not placed so low down relatively to
the mouth of the nectary. The viscid matter
is of a somewhat different chemical nature, as
shown by its much greater viscidity, when
moistened, after having been long dried, or
after being kept in weak spirits. The drum-
like pedicel can hardly be said to be present,
and is represented by a longitudinal ridge,

truncated at the end where the caudicle is
attached, and there is hardly a vestige of the
rudimentary tail of the caudicle. In Fig. XII.
the discs of both species, of the proper proportional sizes, are represented as seen from
vertically above. The

Fig. XII.

pollinia, after removal
from their cells, undergo
the same movements;
but the inward move-
ment seemed to be
greater than in the last
species, in conformity

B. Disc and caudicle of H. chlorantha,
seen from above, with the drum-
like pedicel fore-shortened.
A. Disc and caudicle of Habenaria
bifolia, seen from above.

with the position of the stigma.   In both
forms the movement is well seen by removing,
with a pair of pincers, a pollinium by the
thick end, and holding it motionless under the
microscope, when the plane of the viscid disc
will be seen to move through an angle of at
least 45°. The caudicles of the Lesser Butterfly Orchis are relatively very much shorter
than in the other species; the little packets
of pollen are shorter, whiter, and, in a mature
flower, separate much more readily from each
other. Lastly, the stigmatic surface is differently shaped, being more plainly tripartite,

with two lateral prominences, situated beneath
the viscid discs. These prominences contract
the mouth of the nectary, making it sub-
quadrangular. Hence I cannot doubt that the
Larger and Lesser Butterfly Orchids are distinct species, masked by close external simi-
larity.

As soon as I had examined the Lesser
Butterfly Orchis, I felt convinced, from the
position of the viscid discs, that it would be
fertilised in a different manner from the
Larger Butterfly Orchis ; and now, owing to
the kindness of Mr. F. Bond, I have examined
two moths, namely, Agrotis segetum and
Anaitis plagiata, one with three pollinia, and
the other with five pollinia, attached, not to
the side of the face as in the last species, but
to the base of the proboscis. I may remark
that the pollinia of these two species of Habenaria, when attached to moths, could be distinguished at a glance.

We have now finished with the Ophreæ ;
but before passing on to the next Division, I
will recapitulate the chief facts on the movements of the pollinia, all due to the nicely

regulated contraction of that small portion of membrane (together with the pedicel in the case of Habenaria) lying between the layer or ball of adhesive matter and the extremity of the caudicle. In most of the species of Orchis the stigma lies directly beneath the anther-cells, and the pollinia simply move vertically downwards. In Orchis pyramidalis and in Gymnadenia there are two lateral and inferior stigmas, and the pollinia move downwards and outwards, diverging at the proper angle (by a different mechanism in the two cases), so as to strike the two lateral stigmas; in Habenaria the stigmatic surface lies beneath and between the two widely separated anther-cells, and the pollinia here again move downwards, but, at the same time, converge. A poet might imagine, that whilst the pollinia are borne from flower to flower through the air, adhering to a moth's body, they voluntarily and eagerly place themselves, in each case, in that exact position in which alone they can hope to gain their wish and perpetuate their race.

CHAPTER III.

Epipactis palustris; curious shape of the labellum and its apparent importance in the fructification of the flower — Cephalanthera grandiflora; rostellum aborted; early penetration of the pollen-tubes; case of imperfect self-fertilisation; fertilisation aided by insects — Goodyera repens — Spiranthes autumnalis; perfect adaptation by which the pollen of a younger flower is carried to the stigma of an older flower on another plant.

WE now come to another great tribe of British Orchids, the Neotteæ, which have a free anther standing behind the stigma: their pollen-grains are tied together by fine elastic threads, which partially cohere and project at the *upper* end of the pollen-mass, and are attached (with some exceptions) to the back of the rostellum. Consequently the pollen-masses have no true and distinct caudicles. In one genus alone (Goodyera) the pollen-grains are collected into packets as in Orchis. Epipactis and Goodyera agree pretty closely in their manner of fertilisation with the Ophreæ, but are more simply organised: Spiranthes comes under the same category, but in some respects is differently modified. Cephalanthera seems to be a de-

Fig. XIII.

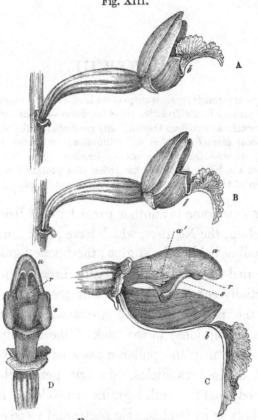

EPIPACTIS PALUSTRIS.

a. anther, with the two open cells seen in the front view D.	auricle, referred to in a future chapter.
a′. rudimentary anther, or	*r.* rostellum. *s.* stigma.
	l. labellum.

A. Side view of flower (with the lower sepals alone removed) in
    its natural position.   B. Side

graded or simplified Epipactis; as it does not possess a rostellum—that eminently characteristic organ—and as its pollen-grains are single, it bears almost the same sort of relation to other Orchids, which a wingless bird does to other birds.

Epipactis palustris.*—The lower part of the stigma is bilobed and projects in front of the column (see *s* in the side and front views, C, D, Fig. XIII.). On its square summit a single, small, nearly globular rostellum is seated. The anterior face of the rostellum (*r*, C, D) projects a little beyond the surface of the upper part of the stigma, and this is of importance. In the early bud the rostellum consists of a friable mass of cells, with the exterior surface rough :

B. Side view of flower, with the distal portion of the labellum depressed, as if by the weight of an insect.

C. Side view of flower, with all the sepals and petals removed, excepting the labellum, of which the near side has been cut away ; the massive anther is seen to be of large size.

D. Front view of column, with all the sepals and petals removed : the rostellum has shrunk down a little in the specimen figured, and ought to have stood higher, so as to hide more of the anther-cells.

---

* I am much indebted to Mr. A. G. More, of Bembridge in the Isle of Wight, for repeatedly sending me fresh specimens of this beautiful Orchis.

these superficial cells undergo a great change
during development, and become converted
into a soft, smooth, highly elastic membrane
or tissue, so excessively tender that it can be
penetrated by a human hair; when thus penetrated, or when slightly rubbed, the surface
becomes milky and in some degree viscid, so
that the pollen-grains adhere to it. In some
cases, though I observed this more plainly in
E. latifolia, the surface of the rostellum apparently becomes milky and viscid without having been touched. This exterior soft elastic
membrane forms a cap to the rostellum, and
is internally lined with a layer of much more
adhesive matter; this matter, when exposed to
the air, dries in from five to ten minutes. By
a slight upward and backward push with any
object, the whole cap, with its viscid lining,
is removed with the greatest ease; and a
minute square stump, the basis of the rostellum, alone is left on the summit of the
stigma.

In the bud-state the anther stands quite free
behind the rostellum and stigma; it opens
longitudinally whilst the flower is still unexpanded, and exposes the two oval pollen-masses,

which are now loose in their cells. The pollen
consists of spherical granules, cohering in
fours, but not affecting each other's shapes;
and these compound grains are tied together
by fine elastic threads. These threads are
collected into bundles, extending longitudinally
along the middle line of the front face of each
pollinium, where it comes into contact with the
back of the uppermost part of the rostellum.
From the number of these threads this middle
line looks brown, and each pollen-mass here
shows a tendency to be divided longitudinally
into two halves : in all these respects there is
a close general resemblance to the pollinia of
the Ophreæ.

The line where the parallel threads are the
most numerous is the line of greatest strength;
elsewhere the pollinia are extremely friable,
so that masses of pollen can be easily broken
off. In the bud-state the rostellum is curved a
little backwards, and is pressed against the
recently-opened anther ; and the above-mentioned slightly projecting bundles of threads
become firmly attached to the posterior flap
of the membranous cap of the rostellum.
The point of attachment lies a little beneath

F

the summit of the pollen-masses ; but the
exact point is somewhat variable, for I have
met with specimens in which the attachment
was one-fifth of the length of the pollen-
masses from their summits. This variability is
so far interesting, as it is a step leading to the
structure of the Ophreæ, in which the confluent
threads, or caudicles, spring from the lower
ends of the pollen-masses. After the pollinia
are firmly attached by their threads to the
back of the rostellum, the rostellum bends
a little forwards, and this partly draws the
pollinia out of their anther-cells. The upper
end of the anther consists of a blunt, solid
point, not including pollen ; this blunt point
projects slightly beyond the face of the rostellum, which circumstance, as we shall see, is
important.

The flowers stand out (Fig. A) horizontally
from the stem. The labellum is curiously
shaped, as may be seen in the drawings : the
distal half, which projects beyond the other
petals and forms an excellent landing-place for
insects, is joined to the basal half by a narrow
hinge, and naturally (Fig. A) is turned a little
upwards, so that its edges pass within the

edges of the basal portion. So flexible and
elastic is the hinge that the weight of even
a fly, as Mr. More informs me, depresses the
distal portion; it is represented in Fig. B in
this state; but when the weight is removed it
instantly springs up to its former (Fig. A) and
ordinary position, and with its curious medial
ridges partly closes the entrance into the flower.
The basal portion of the labellum forms a cup,
which at the proper time is filled with nectar.

Now let us see how all the parts, which I
have been obliged to describe in detail, act.
When I first examined these flowers I was
much perplexed: trying in the same way as I
should have done with a true Orchis, I slightly
pushed the protuberant rostellum downwards,
and it was very easily ruptured; some of the
viscid matter was withdrawn, but the pollinia
remained in their cells. Reflecting on the
structure of the flower, it occurred to me that
an insect in entering to suck the nectar, from
depressing the distal portion of the labellum,
would not touch the rostellum; but that, when
within the flower, from the springing up of
this distal half of the labellum, it would be
almost compelled to back out parallel to the

stigma by the higher part of the flower. I
then brushed the rostellum lightly upwards
and backwards with the end of a feather and
other such objects; and it was pretty to see
how easily the membranous cap of the rostellum
came off, and how well, from its great elasticity, it fitted the object, whatever its shape
might be, and how firmly it clung to it from
the viscidity of its under surface. Together
with the cap large masses of pollen, adhering
by the threads, were necessarily withdrawn.

Nevertheless the pollen-masses were not
nearly so cleanly removed as those which had
been naturally removed by insects. I tried
dozens of flowers, always with the same imperfect result. It then occurred to me that an
insect in backing out of the flower would naturally push with some part of its body against
the blunt and projecting upper end of the
anther which overhangs the stigmatic surface.
Accordingly I so held the brush that, whilst
brushing upwards against the rostellum, I
pushed against the blunt solid end of the
anther (see Sect. C); this at once eased the
pollinia, and they were withdrawn in an entire
state. At last I understood the mechanism of
the flower.

The large anther stands nearly parallel and behind the stigma (Sect. C), so that the pollinia when withdrawn by an insect would naturally adhere to its body in a position fitted to strike, as soon as it visited another flower, the almost parallel stigmatic surface. Hence we have not here, or in any Neotteæ, that movement of depression so common with the pollinia of the Ophreæ. When an insect with the pollinia attached to its back or head enters another flower, the easy depression of the distal portion of the labellum probably plays an important part; for the pollen-masses are extremely friable, and, if struck against the tips of the petals in the act of entering, much of the pollen would be lost; but as it is, an open gangway is offered, and the viscid stigma, with its lower part protuberant, lying in front, is the first object against which the pollen-masses projecting forwards from the insect's back would naturally strike.*   I did not count the flowers,

* As it is quite possible that I may have overrated the importance of the peculiar structure of the labellum, I asked Mr. A. G. More to remove the distal half of the labellum from some flowers before they had expanded, but I was too late in my application. He was able to try only two flowers, which were situated near the summit of the spike. These flowers

but in one lot of spikes sent me by Mr. More
a large majority of the pollinia had been naturally and cleanly removed by some unknown
insect.

Epipactis latifolia.—This species agrees with
the last in all the foregoing specified details,
excepting that the rostellum projects considerably further beyond the face of the stigma,
and the blunt upper end of the anther projects
less. The viscid matter lining the elastic cap
of the rostellum takes a longer time in setting
dry. The upper petals and sepals are more
widely expanded than in E. palustris : the distal portion of the labellum is smaller, and is
firmly united to the basal portion (Fig. XIV.) ;
so that it is not flexible and elastic : it apparently serves only as a landing-place for insects. The fertilisation of this species depends
simply on an insect striking in an upward
and backward direction the highly-protuberant
rostellum, which it would be apt to do in

formed seed-capsules, which were certainly small; but this may
have been owing to their position. Unfortunately also these
capsules shed most of their seed in being sent to me; so that I
could not ascertain whether the seeds were well formed. Of the
few seeds which did remain within these two capsules many were
shrivelled and bad.

retreating after having sucked the copious nectar in the cup of the labellum. Apparently it is not at all necessary that the insect should push back the less protuberant blunt upper end of the anther ; at least I found that the pollinia could be easily removed by simply dragging off the cap of the rostellum in an upward and backward direction.

Fig. XIV.

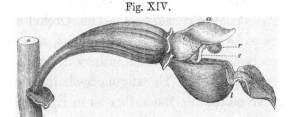

EPIPACTIS LATIFOLIA.

Flower viewed sideways, with all the sepals and petals removed, except the labellum.

a. anther.	*s.* stigma.
r. rostellum.	*l.* labellum.

In Germany C. K. Sprengel caught a fly with the pollinia of this species attached to its back. In England the flowers are much visited by insects : during the wet and cold season of 1860 a friend in Sussex examined five spikes bearing eighty-five expanded flowers : of these, fifty-three had the pollinia removed, and thirty-

two had them in place ; but as many of the
latter were immediately beneath the buds, ultimately a larger proportion would almost certainly have been renewed. In Devonshire I
found a spike with nine open flowers, and the
pollinia in all were removed with one exception, in which a fly, too small to remove the
pollinia, had become glued to it and to the
stigma, and had there miserably perished.

Cephalanthera grandiflora.—This Orchid appears to be closely allied to Epipactis, though
it has been ranked by some authors in a widely
different position. The stigma holds the same
relative position to the anther as in Epipactis ;
but we have here the unique case (my remarks
never apply to the very different group of the
Cypripediæ) of there being no rostellum. The
anther resembles that of Epipactis, but stands
rather higher up relatively to the stigma.
The pollen is extremely friable, and readily
adheres to any object rubbed against it ; the
spherical grains are separate, instead of being
united by threes or fours, as in all other
Orchids ;* they are tied together by a few

---

* This separation of the grains was observed, and is represented,
by Bauer in the plate published by Lindley in his magnificent
' Illustrations of Orchidaceous Plants.'

Fig. XV.

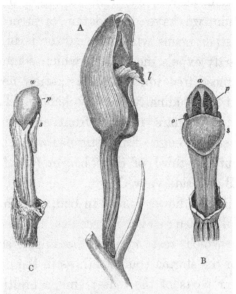

CEPHALANTHERA GRANDIFLORA.

a. anther ; in the front view, B, the two cells with the pollen are seen.	*p.* masses of pollen.
	s. stigma.
o. rudimentary anther, or auricle.	*l.* distal portion of the labellum.

A. Oblique view of perfect flower, when fully expanded.
B. Front view of column, with all the sepals and petals removed.
C. Side view of column, with all the sepals and petals removed ; the narrow pillars of pollen between the anther and stigma can just be seen.

weak elastic threads; so that in the state of
the pollen, as well as in the abortion of the
rostellum, we have degradation of structure.
The anther opens whilst the flower is in bud
and partly expels the pollen, which stands in
two almost free upright pillars, each nearly
divided longitudinally into two halves. These
subdivided pillars rest in front against the
upper square edge of the stigma, which rises
to about one-third of their height (see front
view B and side view C).

Whilst the flower is still in bud, or before it
is as fully open as ever it becomes, the pollen-
grains which rest against the upper sharp
edge of the stigma (but not those in the upper
or lower parts of the mass) emit a multitude
of tubes, deeply penetrating the stigmatic
tissue. After this period the stigma bends a
little forward, and the result is that the two
friable pillars of pollen stand almost free from
their anther-cells, being tied to and supported
in front by the penetration of their pollen-tubes
into the edge of the stigma. Without this
support the pillars would soon fall down.

Differently from Epipactis, the flower stands
upright; the lower part of the labellum is

turned up parallel to the column (Fig. A), and the tips of the lateral petals never become quite separate ; * so that the pillars of pollen are protected from the wind, and as the flower stands upright they do not tumble down from their weight. These are points of high importance to the plant, as the pollen would otherwise be blown or fall down and be wasted. The labellum is formed of two portions, as in Epipactis ; and when the flower is mature, the small triangular distal portion turns down at right angles to the basal portion; thus forming a small landing-place in front of a triangular door, situated half-way up the almost tubular flower, by which insects can enter. I did not observe any nectar ; but as the lower cup-formed portion of the labellum presents the same structure as in Epipactis, I presume it is secreted. After a short time, as soon as the flower is fully fertilised, the small distal portion of the labellum rises up, shuts the triangular door, and again perfectly encloses the organs of fructification.

* Bauer figures the flowers much more widely expanded : all that I can say is that I have not seen them in this condition.

In the early penetration of the stigma by a
multitude of pollen-tubes, which I traced far
down the stigmatic tissue, we apparently have
another case, like that of the Bee Ophrys, of
perpetual self-fertilisation. I was much surprised at this fact, and asked myself: Why
does the distal portion of the labellum open for
a short period? what is the use of the great
mass of pollen above and below that layer
of grains, the tubes of which alone penetrate
the upper edge of the stigma? The stigma
has a large flat viscid surface; and during
several years I have almost invariably found
masses of pollen adhering to its surface, and
the friable pillars of pollen by some means
broken down. It occurred to me that,
although the flowers stand upright, and the
pillars are well protected from the wind,
yet that the pollen-masses might ultimately
topple over from their own weight, and so
fall on the stigma, thus completing the act
of self-fertilisation. Accordingly, I covered
up with a net a plant with four buds, and
examined the flowers as soon as they had
withered; the broad stigmas of three of these
flowers were perfectly free from pollen, but

a little had fallen on one corner of the fourth. With the exception, also, of the summit of the pillar of pollen in this latter flower, all the other pillars still stood upright and unbroken. I looked at the flowers of some surrounding plants, and everywhere found, as I had so often done before, broken-down pillars and masses of pollen on the stigmas.

Hence we may safely infer that insects visit the flowers, disturb the pollen, and leave masses of it on the stigmas. We thus see that the reflexion of the distal portion of the labellum, by which a temporary landing-place and door are formed,—the upturned labellum, by which the flower is made tubular and insects are compelled to crawl close by the stigmatic surface,—the pollen readily cohering to any object, and standing in friable pillars protected from the wind,—and, lastly, the large masses of pollen standing above and below that layer of pollen-grains, of which alone the tubes penetrate the edge of the stigma,—are all co-ordinated structures, far from useless; and useless they would be if these flowers were capable of perfect self-fertilisation.

To ascertain how far the early and invariable penetration of the upper edge of the stigma by the pollen-grains, which rest on it, is effectual for fertilisation, I covered up a plant, just before the flowers opened, and removed the thin net as soon as they had begun to wither. From long experience I am sure that this temporary covering could not have injured their fertility. The four covered flowers produced as fine seed-capsules as any on the surrounding plants. When ripe, I gathered them, and likewise capsules from several surrounding plants, growing under similar conditions, and weighed the seed in a chemical balance. The seed from the four capsules of the uncovered plants weighed 1·5 grain, from the covered plant the seed of an equal number of capsules weighed under one grain; but this does not give a fair idea of the relative difference of fertility, for I observed that a great number of the seeds from the covered plant were mere minute and shrivelled husks. Accordingly I mixed the seeds well together, and took four little lots from one heap and four little lots from the other heap, and, having soaked them in

water, compared them under the compound
microscope : out of forty seeds from the
uncovered plants there were only four bad,
whereas of forty seeds from the covered-up
plants there were at least twenty-seven bad;
so that there were nearly seven times as
many bad seeds in the covered-up plants as
in those left to the free access of insects.

Hence we have the following curious and
complex case. Perpetual self-fertilisation,
but in an extremely imperfect degree, by the
early penetration of the pollen-tubes; this
would be useful to the plant, if insects failed
to visit the flowers. The penetration of the
pollen-tubes, however, is apparently more
important in retaining the pillars of pollen
in their proper places, so that insects, in
crawling into the flowers, might get dusted
with pollen. This imperfect self-fertilisation
is habitually and largely aided by insects,
which may carry a flower's own pollen on
to its stigma ; but an insect thus smeared
with pollen could not fail likewise to cross
distinct individuals. From the relative position of the parts, it seems indeed probable
(but I omitted to prove this by the early

removal of the anthers, so as to observe
whether pollen was brought to the stigma
from other flowers) that an insect would
more frequently get dusted by crawling out
of a flower than by crawling into one; and
this would of course facilitate a union between two distinct individuals. Hence Cephalanthera offers only a partial exception to
the almost universal rule that the flowers of
Orchids are fertilised by the pollen of other
flowers.

Goodyera repens.*—This genus, in most of
its characters with which we are concerned,
is rather closely related to Epipactis. The
shield-like rostellum is almost square, and
projects beyond the stigma; it is supported on
each side by sloping sides rising from the upper
edge of the stigma, in nearly the same manner
as we shall presently see in Spiranthes. The
surface of the protuberant part of the rostellum
is rough, and when dry can be seen to be
formed of cells; it is delicate, and, when slightly
pricked, exudes a little milky viscid fluid; it is
lined by a layer of very adhesive matter, which

---

* Specimens of this rare Highland Orchid were most kindly
sent me by the Rev. G. Gordon of Elgin.

quickly sets hard when exposed to the air. The protuberant surface of the rostellum, when gently rubbed upwards, is easily removed, and carries with it a strip of membrane, to the hind part of which the pollinia are attached. The sloping sides which support the rostellum are not removed at the same time, but remain projecting up like a fork and soon wither. The anther is borne on an elongated broad filament; a membrane on both sides unites this filament to the edges of the stigma, forming an imperfect cup or clinandrum. The anther-cells open in the bud, and the pollinia become attached by their anterior faces, just beneath their summits, to the back of the rostellum; ultimately the anther opens widely, leaving the pollinia almost naked, but partially protected within the membranous cup or clinandrum. Each pollinium is partially divided lengthways; the pollen-grains cohere in subtriangular packets, including a multitude of grains, each grain consisting of four granules; and these packets are tied together by strong elastic threads, which at their upper ends run together and form a single flattened brown elastic ribbon, of which the truncated extremity adheres to the back of the rostellum.

The surface of the orbicular stigma is remarkably viscid, which is necessary in order that the unusually strong threads connecting the packets of pollen should be ruptured. The labellum is partially divided into two portions; the terminal portion is reflexed, and the basal portion is cup-formed and filled with nectar. The passage into the flower between the rostellum and labellum is contracted. Since my examination of Spiranthes, immediately to be described, I have suspected that the labellum moves further from the column in mature flowers, in order to allow insects, with the pollinia adhering to their heads or probosces, to enter the flowers more freely. In many of the specimens received, the pollinia had been removed by insects, and the fork-shaped supporting sides of the rostellum were partially withered.

Goodyera is an interesting connecting link between several very distinct forms. In no other member of the Neotteæ have I seen so near an approach to the formation of a true caudicle,* like that in the Ophreæ; and it

---

* In a foreign species, the Goodyera discolor, sent me by Mr. Bateman, the pollinia approach in structure still closer to those of the Ophreæ; for the pollinia thin out into long caudicles, closely resembling in form those of an Orchis. The caudicle is

is curious that in this genus alone (as far as I have seen) the pollen-grains cohere in large packets, as in the Ophreæ. If the nascent caudicles had been attached to the lower ends of the pollinia, and they are attached a little beneath the summit, the pollinia would have been almost identical with those of a true Orchis. In the rostellum being supported by sloping sides, which wither when the viscid disc is removed,—in the membranous cup or clinandrum between the stigma and anther,—and in some other respects, we have a clear affinity to Spiranthes. In the anther having a broad filament we see a relation to Cephalanthera. In the structure of the rostellum, with the exception of the sloping sides, and in the shape of the labellum, we see the affinity of Goodyera to Epipactis. Goodyera probably shows us the state of the organs of fructifica-

formed of a bundle of elastic threads, with very small and thin packets of pollen-grains attached to them, and arranged like tiles one over the other. The two caudicles are united together near their bases, where they are attached to a disc of membrane lined with viscid matter. From the small size and extreme thinness of the basal packets of pollen, and from the strength of their attachment to the threads, I believe that they are in a functionless condition; if so, these prolongations of the pollinia are true caudicles.

tion in a large group of Orchids, now mostly extinct, but the parents of many living descendants.

Spiranthes autumnalis.—This Orchid, with its pretty name of Ladies'-tresses, presents some interesting peculiarities.* The rostellum is a long, thin, flat, projection, joined by sloping shoulders to the summit of the stigma. In the middle of the rostellum a narrow vertical brown object (Fig. XVI., C) may be seen, bordered on each side and covered by transparent membrane. This brown object I will call "the boat-formed disc." It forms the middle portion of the posterior surface of the rostellum, and consists of a narrow strip of the exterior membrane in a modified condition. Its summit (Fig. E) is pointed, its lower end rounded, and it is slightly bowed, so as altogether to resemble a boat or canoe. It is rather more than $\frac{4}{100}$ of an inch in length, and less than $\frac{1}{100}$ in breadth. It is nearly rigid, and appears fibrous, but is really formed of elongated and thickened cells, partially confluent.

---

* I am much indebted to Dr. Battersby of Torquay, and to Mr. A. G. More of Bembridge, for sending me specimens; but I have subsequently examined many growing plants.

Fig. XVI.

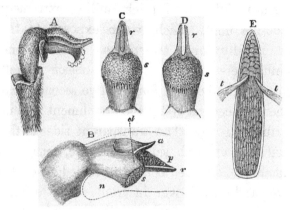

SPIRANTHES AUTUMNALIS, OR LADIES' TRESSES.

a. anther.	cl. margin of clinandrum.
p. pollen-grains.	r. rostellum.
t. threads of the pollen-	s. stigma.
masses.	n. nectar-receptacle.

A. Side view of flower in its natural position, with the two lower sepals alone removed. The labellum can be recognised by its fringed and reflexed lip.

B. Side view of a mature flower, with all the sepals and petals removed. The position of the labellum (which has moved from the rostellum) and of the upper sepal is shown by the dotted lines.

C. Front view of the stigma, and of the rostellum with its embedded central disc.

D. Front view of the stigma and of the rostellum after the viscid disc has been removed.

E. Viscid disc, removed from the rostellum, greatly magnified, viewed posteriorly, and with the attached elastic threads of the pollen-masses; the pollen-grains have been removed from the threads.

This boat, standing vertically up on its stern, is filled with thick, milky, extremely adhesive fluid, which, when exposed to the air, rapidly turns brown, and in about one minute sets quite hard. An object is well glued to the boat in four or five seconds, and when the cement is dry the attachment is wonderfully strong. The transparent sides of the rostellum, on each side of the disc, consist of membrane, attached behind to the edges of the boat, and folded over in front, so as to form the anterior face of the rostellum. This folded membrane, therefore, covers, almost like a deck, the cargo of viscid matter within the boat.

The anterior face of the rostellum is slightly furrowed in a longitudinal line over the middle of the boat, and is endowed with a remarkable vital property; for, if the furrow be touched very gently by a needle, or if a bristle be laid along the furrow, it instantly splits along its whole length, and a little milky adhesive fluid exudes. This action is not mechanical, or due to simple violence. The fissure runs up the whole length of the rostellum, from the stigma beneath to the summit: at the summit the fissure bifurcates,

and runs down the back of the rostellum on
each side and round the stern of the boat-
formed disc.   Hence after this splitting action
the boat-formed disc lies quite free, but embedded in a fork in the rostellum.   The act of
splitting apparently never takes place spontaneously.   I covered up under a net a plant
with unexpanded flowers, and five of these
flowers remained fully expanded for exactly a
week under the net: I then examined their
rostella, and not one had split; whereas almost
every flower on the surrounding and uncovered spikes, which had been visited and
touched by insects, after remaining expanded
for only twenty-four hours, had their rostella
fissured.   Exposure for two minutes to the
vapour of a little chloroform causes the rostellum to split; and this we shall hereafter see
is likewise the case with some other Orchids.

When a bristle is laid for two or three
seconds in the furrow of the rostellum, and
the membrane has consequently become fissured, the viscid matter within the boat-formed
disc lies so close to the surface, and, indeed,
slightly exudes, that the disc is almost sure
to be glued longitudinally to the bristle and

to be withdrawn with it. When the disc is
withdrawn the two sides of the rostellum
(Fig. D), which have been described by
some botanists as two distinct foliaceous projections, are left sticking up like a fork. This
is the common condition of the flowers after
they have been open for two or three days,
and have been visited by insects. The fork
soon withers.

When the flower is in bud the back of the
boat-formed disc is covered with a layer of
large rounded cells, so that the disc does not
strictly form the exterior surface of the back
of the rostellum. These cells contain slightly
viscid matter : they remain unaltered (as may
be seen at Fig. E) towards the apex of the
disc, but at the point where the pollinia are
attached they disappear. Hence I concluded
that the viscid matter contained in these
cells, when they burst, served to fasten the
threads of the pollinia to the disc ; but, as
in several large exotic Orchids I could see no
trace of such cells, I presume that this view is
erroneous.

The stigma lies beneath the rostellum, and
projects with a sloping surface (see side-view,

B) ; its lower margin is rounded and fringed
with hairs. On each side a membrane (c l, B)
extends from the edges of the stigma to the
filament of the anther, thus forming a membranous cup or clinandrum, in which the
lower ends of the pollen-masses lie protected.
Each pollinium consists of two leaves of
pollen, quite disconnected at their lower ends,
and with their summits distinct, but they are
united by elastic threads for about half of
their length : a very slight modification would
convert the two pollinia into four leaves of
pollen, as occurs in the genus Malaxis and
in many foreign Orchids. Each of the four
leaves, moreover, consists of a double layer of
pollen-grains, united only along their edges.
The pollen-grains (each consisting of four
granules) are united by elastic threads, which
are more numerous along the edges of the
leaves, and converge at the summit of each
pollinium. The leaves of pollen are very
brittle, and, when placed on the viscid stigma
of a flower, large sheets are easily broken off.
  Long before the flower expands, the anther-
cells, which are pressed against the back of
the rostellum, open in their upper part, so that

the included pollinia come exactly into contact
with the back of the boat-formed disc; the
projecting threads then become firmly attached
to rather above the middle part of the back
of the disc. The anther-cells afterwards
open lower down, and their membranous
walls contract and become brown; so that by
the time the flower has fully expanded the
upper parts of the pollinia lie quite naked,
their bases rest in little cups formed by the
withered anther-cells, and they are laterally
protected by the clinandrum. As the pollinia
thus lie loose, they are easily removed.

The tubular flowers are elegantly arranged
in a spire round the spike, extending horizontally (Fig. A) from it. The labellum is
channelled down the middle, and is furnished
with a reflexed and fringed lip, on which bees
alight; its basal internal angles are produced
into two globular processes, which secrete an
abundance of nectar. The nectar is collected
(n, Fig. B) in a small receptacle beneath.
Owing to the protuberance of the lower
margin of the stigma, and of the two lateral
inflexed nectaries, the orifice into the nectar-
receptacle is much contracted, and is central.

When the flower first opens the receptacle contains nectar, and at this period the front of the rostellum, which is slightly furrowed, lies close to the channelled labellum; consequently a passage is left, but so narrow that only a fine bristle can be passed down it.* In a day or two the labellum moves a little farther from the rostellum, and a wider passage is left to the stigmatic surface. On this slight movement of the labellum the fertilisation of the flower absolutely depends.

With most Orchids the flowers remain open for some time before they are visited by insects; but with Spiranthes I have generally found the boat-formed disc removed very soon after the expansion of the flower. For example, in the two last spikes which I happened to examine there were in one numerous buds on the summit, with the seven lowest flowers alone expanded, of which six had their discs and pollinia removed; the other spike had eight expanded flowers, with the pollinia of all removed. We have seen

* Professor Asa Gray was so kind as to examine Spiranthes gracilis and cernua in the United States. He found the same general structure as in our S. autumnalis, and he was struck with the narrowness of the passage into the flower.

that when the flower first opens it would be
attractive to insects, for the receptacle already
contains nectar ; and the rostellum at this
period lies so close to the channelled labellum
that a bee or moth could not pass down its
proboscis without touching the medial furrow
of the rostellum.   This I know to be the case
by repeated trials with a bristle.

Let it be observed how beautifully every-
thing is contrived that the pollinia should be
withdrawn by an insect visiting the flower.
The pollinia are already attached to the disc
by their threads, and, from the early withering
of the anther-cells, they hang loosely suspended but protected within the clinandrum.
The touch of the proboscis causes the rostellum
to split in front and behind, and frees the
long, narrow, boat-formed disc, which is loaded
with extremely viscid matter, sure to adhere
longitudinally to the proboscis.   When the
bee flies away, so surely will it carry away
the pollinia.   As the pollinia are attached
parallel to the disc, they will become attached
parallel to the proboscis.   Here, however,
appears a difficulty ; when the flower first
opens and is best adapted for the removal of

the pollinia, the labellum lies so close to the
rostellum, that the pollinia, when attached to
the proboscis of an insect, could not possibly be
forced into the flower so as to reach the stigma ;
they would be either upturned or broken off ;
but we have seen that after two or three
days the labellum becomes more reflexed and
moves from the rostellum,—a wider passage
being thus left. When in this condition, I
have tried with the pollinia attached to a
fine bristle ; and by inserting the bristle into
the nectar-receptacle (n, Fig. B), it is pretty
to see how sheets of pollen are left adhering
to the viscid stigma. It may be observed in
the diagram, B, that the orifice into the
nectar-receptacle lies, owing to the projection
of the stigma, close to the lower side of the
flower ; hence insects would insert their
probosces on this side, and an open space is
thus afforded for the attached pollinia to be
carried down to the stigma, without being
brushed off. The stigma evidently projects
so that the ends of the pollinia may strike
against it.

Hence, in Spiranthes, not only the pollen
must be carried from one flower to another, as

in most Orchids, but a lately expanded flower, which has its pollinia in the best state for removal, cannot then be fertilised. Generally old flowers will be fertilised by the pollen of younger flowers, borne, as we shall see, on a separate plant. In conformity with this I observed that the stigmatic surfaces of the older flowers were far more viscid than those of the younger flowers. Nevertheless, a flower which in its early state had not been visited by insects would not necessarily, in its later and more expanded condition, have its pollen wasted; for insects, in inserting and withdrawing their probosces, bow them forwards, and would thus often strike the furrow in the rostellum. I imitated this action with a bristle, and often succeeded in withdrawing the pollinia. I was led to try this from at first choosing older flowers for examination; and passing a bristle, or fine culm of grass, straight down into the nectary, the pollinia were never withdrawn; but when I bowed it forward, I succeeded. Those flowers which have not their own pollinia removed can of course be fertilised, and I have seen not a few cases of flowers with their pollinia still in place, with sheets of pollen on their stigmas.

At Torquay I watched a number of these flowers growing together for about half an hour, and saw three humble-bees of two kinds visit them. I caught one and examined its proboscis : on the superior lamina, some little way from the tip, two perfect pollinia were attached, and three other boat-formed discs without pollen ; so that this bee had removed the pollinia from five flowers, and had probably left the pollen of three of them on the stigmas of other flowers. The next day I watched the same flowers for a quarter of an hour, and caught another humble-bee at work ; one perfect pollinium and four boat-formed discs adhered to its proboscis, one on the top of the other, showing how exactly the same part had each time touched the rostellum.

The bees always alighted at the bottom of the spike, and, crawling spirally up it, sucked one flower after the other. I believe humble-bees generally act thus when visiting a dense spike of flowers, as it is most convenient for them ; in the same manner as a woodpecker

always climbs up a tree in search of insects.
This seems a most insignificant observation ;
but see the result. In the early morning,
when the bee starts on her rounds, let us suppose that she alighted on the summit of the
spike ; she would surely extract the pollinia
from the uppermost and last opened flowers ;
but when visiting the next succeeding flower,
of which the labellum in all probability would
not as yet have moved from the column (for
this is slowly and very gradually effected),
the pollen-masses would often be brushed off
her proboscis and be wasted. But nature
suffers no such waste. The bee goes first to
the lowest flower, and, crawling spirally up
the spike, effects nothing on the first spike
which she visits till she reaches the upper
flowers, then she withdraws the pollinia : she
soon flies to another plant, and, alighting on
the lowest and oldest flower, into which there
will be a wide passage from the greater reflexion of the labellum, the pollinia will strike
the protuberant stigma : if the stigma of the
lowest flower has already been fully fertilised,
little or no pollen will be left on its dried
surface ; but on the next succeeding flower,

of which the stigma is viscid, large sheets of pollen will be left. Then as soon as the bee arrives near the summit of the spike she will again withdraw fresh pollinia, will fly to the lower flowers on another plant, and fertilise them ; and thus, as she goes her rounds and adds to her store of honey, she will continually fertilise fresh flowers and perpetuate the race of our autumnal Spiranthes, which will yield honey to future generations of bees.

CHAPTER IV.

Malaxis paludosa; simple means of fertilisation — Listera ovata; sensitiveness of the rostellum; explosion of viscid matter action of insects; perfect adaptation of the several organs — Listera cordata — Neottia nidus-avis; its fertilisation effected in the same manner as in Listera.

WE now come to the last of the British Orchids, those in which no portion of the exterior membranous surface of the rostellum is permanently attached to the pollinia. This subdivision includes only three genera known to me, namely, Malaxis, Listera, and Neottia, which are here grouped together merely for convenience. Malaxis in its fertilisation is an uninteresting form; Listera and Neottia, on the other hand, are amongst the most remarkable of all Orchids from the manner in which their pollinia are removed by insects, through the sudden explosion of viscid matter contained within their rostellums.

Malaxis paludosa.—This rare Orchid,* the least of the British species, differs from the

* I am greatly indebted to Mr. Wallis, of Hartfield, in Sussex, for numerous living specimens of this Orchid.

others in the position of its flowers. In Malaxis the labellum is turned upwards,* instead of downwards, so as to afford, as in other Orchids, a landing-place for insects; its lower margin clasps the column, making the entrance into the flower tubular. From its position it partially protects the organs of fructification. [Fig. XVII.] In most Orchids the upper sepal and the two upper petals afford protection; but here they are reflexed in a singular manner (as may be seen in the drawing, Fig. A), apparently to allow insects freely to visit the flower. This position of the flower is the more remarkable, because it has been purposely acquired, as shown by the ovarium being spirally twisted. In all Orchids the labellum is properly directed upwards, but assumes its usual position as the *lower lip* by the twisting of the ovarium; but in Malaxis the twisting has been carried to that degree that the flower occupies the position which it would have held if the ovarium

* Sir James Smith, I believe, first noticed this fact in the 'English Flora,' vol. iv. p. 47, 1828. Towards the summit of the spike the upper sepal does not depend, as represented in the woodcut (Fig. A), but sticks out nearly at right angles; nor is the flower always so completely twisted round.

Fig. XVII.

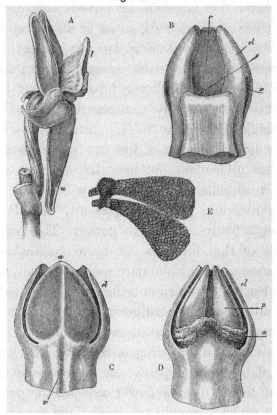

MALAXIS PALUDOSA.

(Partly copied from Bauer, but modified from living specimens.)

a. anther.	*r.* rostellum.
p. pollen.	*s.* stigma.
cl. clinandrum.	*l.* labellum.
v. spiral vessels.	*u.* upper sepal.

A. Perfect

had not been at all twisted, and which the ovarium ultimately assumes when ripe, by a process of gradual untwisting.

When the minute flower is dissected, the column is seen to be longitudinally tripartite; the middle portion of the upper half (see Fig. B) is the rostellum. The upper edge of the lower part of the column projects where united to the base of the rostellum, and forms a rather deep fold. This fold is the stigmatic cavity, and may be compared to a waistcoat-pocket. I found pollen-masses which had their broad ends pushed by insects into this pocket; and a bundle of pollen-tubes had here penetrated the stigmatic tissue.

The rostellum, or middle portion, is a tall membranous projection of a whitish colour,

---

A. Perfect flower viewed laterally, with the labellum in its natural position, upwards.

B. Column viewed in front, showing the rostellum, the pocket-like stigma, and the anterior lateral portions of the clinandrum.

C. Back view of the column in a flower-bud, showing the anther with the pear-shaped included pollinia dimly seen, and the posterior edges of the clinandrum.

D. Back view of an expanded flower, with the anther now contracted and shrivelled, exposing the pollinia.

E. The two pollinia attached to a little transverse mass of viscid matter, hardened by spirits of wine.

formed of square cells, and is covered with a
thin layer of viscid matter : it is slightly concave posteriorly, and its crest is surmounted
by a minute tongue-shaped mass of viscid
matter. The column, with its narrow pocket-
like stigma and with the rostellum above, is
united on each side to a green membranous
expansion, convex exteriorly and concave interiorly, of which the summits on each side
are pointed, and stand a little above the crest
of the rostellum. These two membranes sweep
round (see back views, Figs. C and D), and are
united to the filament or base of the anther ;
thus forming a deep cup or clinandrum behind
the rostellum. The use of this cup is to afford
protection, as we shall immediately see, to the
pollen-masses.   When I have to treat of the
homologies of the different parts, it will be
shown by the course of the spiral vessels that
these two membranes, forming the clinandrum,
consist of the two upper anthers of the inner
whorl, in a rudimentary condition, but utilised
for this special purpose.

In a flower before it expands, a little mass
or drop of viscid fluid may be seen on the
crest of the rostellum, rather overhanging its

front surface. After the flower has remained
open for a little time, this drop shrinks and
becomes more viscid. Its chemical nature is
different from that of the viscid matter in
most Orchids, for it remains fluid for many
days, though fully exposed to the air. From
these facts I concluded that the viscid fluid
exuded from the crest of the rostellum ; but
fortunately I examined a closely-allied Indian
form, namely, the Microstylis Rhedii (sent
me from Kew by Dr. Hooker), and in this,
before the flower opened, there was a similar
drop of viscid matter; but on opening a
still younger bud, I found a minute, regular,
tongue-shaped projection on the crest of the
rostellum, formed of cells, which when slightly
disturbed resolved themselves into a drop of
viscid matter. At this age, also, the front
surface of the whole rostellum, between its
crest and the pocket-like stigma, was coated
with cells filled with similar brown viscid
matter ; so that there can be no doubt, had I
examined a young enough bud of Malaxis,
I should have found a similar minute tongue-
shaped cellular projection on the crest of the
rostellum.

The anther opens widely whilst the flower is in bud, and then shrivels and contracts downwards, so that, when the flower is fully expanded, the pollinia are quite naked, with the exception of their broad lower ends, which rest in two little cups formed by the shrivelled anther-cells. This contraction of the anther is represented in Fig. D in comparison with Fig. C, which shows the state of the anther in a bud. The upper and much pointed ends of the pollinia rest on, but project beyond, the crest of the rostellum; in the bud they are unattached, but by the time the flower opens they are always caught by the posterior surface of the drop of viscid matter, of which the anterior surface projects slightly beyond the face of the rostellum; that they are caught without any mechanical aid I ascertained by allowing some buds to open in my room. In Fig. E the pollinia are shown exactly as they appeared, but not quite in their natural position, when removed by a needle from a specimen kept in spirits of wine, in which the irregular little mass of viscid matter had become hardened and adhered firmly to their tips.

The pollinia consist of two pair of very

thin leaves of waxy pollen ; these four leaves
are formed of angular grains (each apparently
subdivided into four granules), which never
separate. As the pollinia are almost loose,
being retained merely by their tips adhering
to the viscid fluid, and by their bases resting
in the shrivelled anther-cells, and as the petals
and sepals are so much reflexed, the pollinia
would be exposed in a remarkable degree
when the flower is fully expanded, and would
be liable to be blown out of their proper position, had it not been for the membranous
expansions on each side of the column forming
the clinandrum, within which they safely lie.

When an insect inserts its proboscis, or
head, into the narrow space between the
upright labellum and the rostellum, it will
infallibly touch the little projecting viscid
mass, and when it flies away it will withdraw
the pollinia, already attached to the viscid
matter, but otherwise loose. I easily imitated
this action by inserting any small object into
the tubular flower between the labellum and
rostellum. When the insect visits another
flower, the very thin pollen-leaves attached
parallel to the proboscis, or head, will be

forced in, and their broad ends will enter the
pocket-like stigma. I found pollinia in this
position glued to the upper membranous expansion of the rostellum, and with a large
number of pollen-tubes penetrating the stigmatic tissue. The use of the thin layer of
viscid matter, which coats the surface of the
rostellum in this genus and in Microstylis, and
which is of no use in the transportal of the
pollen from flower to flower, seems to be to
keep the leaves of pollen, when brought by
insects, in a proper position for entering and
remaining in the narrow stigmatic cavity.
This fact is rather interesting under a homological point of view, for, as we shall hereafter
see, the primordial nature and purpose of
the viscid matter of the rostellum is that
common to the viscid matter on the stigmas
of most flowers, namely, the retention of the
pollen, when brought by any means to its
surface.

The flowers of the Malaxis, though so small
and inconspicuous, are highly attractive to
insects; for the pollinia had been removed
from all the flowers on each spike, excepting
from one or two close under the buds. In

some old flower-spikes every single pollinium
had been removed. Insects sometimes remove only one of the two pairs. I noticed
one flower with all its four pollinia in place,
with a single pollen-leaf within the stigmatic
cavity ; and this must clearly have been
brought by some insect. I found pollen-
leaves on the stigmas of many other flowers.
The plant produces plenty of seed : on one
spike, thirteen of the twenty-one lower flowers
had formed large capsules.

Listera ovata, or Tway-blade.—The structure
and action of the rostellum of this Orchid has
been the subject of a highly remarkable paper
in the Philosophical Transactions, by Dr.
Hooker,* who has described minutely, and of
course correctly, its curious structure ; he did
not, however, attend to the part which insects
play in the fertilisation of this flower. C. K.
Sprengel well knew the importance of insect-
agency, but he misunderstood both the structure and the action of the rostellum.

The rostellum is of large size, thin, or
foliaceous ; convex in front and concave be-

* 'Philosophical Transactions,' 1854, p. 259.

hind, with its sharp summit slightly hollowed
out on each side ; it arches over the stigmatic
surface (Fig. XVIII. *r, s* A).   Internally, according to Dr. Hooker, it is divided by longitudinal septa into a series of loculi, which
contain and subsequently expel with violence
viscid matter.   These loculi show traces of
their original cellular structure.   I have met
with this structure of the rostellum in no other
genus excepting the closely allied Neottia.
The anther, situated behind the rostellum,
and protected by a broad expansion of the
top of the column, opens in the bud.   The
pollinia, when the flower is fully expanded,
are left quite free, supported behind by the
anther-cells, and lying in front against the
concave back of the rostellum, with their
upper and pointed ends resting on its crest.
Each pollinium is almost divided into two
masses.   The pollen-grains are attached together in the usual manner by a few elastic
threads ; but the threads are weak, and large
masses of pollen can be easily broken off.
After the flower has long remained open,
the pollen becomes more friable.   The labellum is much elongated, contracted at the

Fig. XVIII.

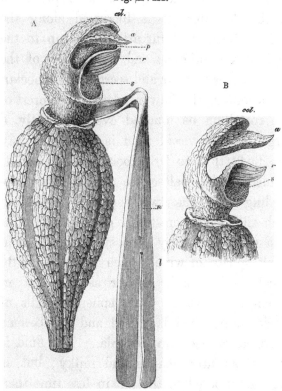

Listera ovata, or Tway-blade (partly copied from Hooker).

col. summit of column.	*s.* stigma.
a. anther.	*l.* labellum.
p. pollen.	*n.* nectar-secreting furrow.
r. rostellum.	

A. Flower viewed laterally, with all the sepals and petals, except the labellum, removed.

B. Ditto, with the pollinia removed, and with the rostellum more reflexed after its explosion.

base, and bent downwards, as represented
in the drawing; it is furrowed along the
middle, from its bifurcation close up to the
base of the stigma; and the borders of the
furrow are glandular and secrete much nectar.

As soon as the flower opens, if the crest of
the rostellum be touched ever so lightly, a
large drop of viscid fluid is instantaneously
expelled; this, as Dr. Hooker has shown, is
formed by the coalescence of two drops proceeding from the two depressed spaces on
each side of the crest. A good proof of this
fact was afforded by some specimens kept in
weak spirits of wine, which had apparently
expelled the viscid matter slowly, and in
which two separate little spherical balls of
viscid matter had hardened and had become
attached to the two pollinia. The fluid is
at first slightly opaque and milky; but as
soon as exposed to the air, in less time than
a second, a film is formed over it, and in two
or three seconds the whole drop sets hard,
and soon assumes a purplish-brown tint. So
exquisitely sensitive is the rostellum, that a
touch from the thinnest human hair suffices
to cause the explosion. It will take place

under water. Exposure to the vapour of chloroform for about one minute also causes the explosion. The viscid fluid when pressed between two plates of glass before it has set hard is seen to be structureless ; but it assumes a reticulated appearance, perhaps caused by the presence and union of globules of a denser fluid immersed in a thinner fluid. As the pointed tips of the loose pollinia lie on the crest of the rostellum, they are always caught by the exploded drop : I have never once seen this fail. So rapid is the explosion and so viscid the fluid, that it is difficult to touch the rostellum with a needle quickly enough not to catch the pollinia already attached to the partially-hardened drop. Hence, if a bunch of flowers be carried home in the hand, some of the sepals or petals, being shaken, will be almost sure to touch the rostellum and withdraw the pollinia, which will cause the false appearance of their having been violently ejected to a distance.

After the anther-cells have opened and have left the naked pollinia resting on the concave back of the rostellum, the rostellum curves a little forwards, and perhaps the anther also moves a little backwards. This movement is of much importance, as the tips of the anther would otherwise be caught by the exploded viscid matter, and the pollinia would for ever be locked up and rendered useless. I once found an injured flower which had been pressed and had exploded before fully expanding, and the anther, with the enclosed pollen-masses, was permanently glued down to the crest of the rostellum. At the moment of the explosion the rostellum, which already stands arched over the stigma, quickly bends forwards and downwards so as to stand (Fig. B) at right angles to its surface. The pollinia, if not removed by the touching object that causes the explosion, become fixed to the rostellum, and by its movement are drawn a little forward. If their lower ends are now freed by a needle from the anther-cells, they spring up; but they are not by this movement placed on the stigma. In the course of some hours, or of a day, the rostellum not only slowly recovers its original slightly-arched position, but becomes quite straight and parallel to the stigmatic surface. This backward movement of the rostellum is of importance; for if after the explosion it permanently remained projecting at right angles close over the stigma, pollen could with difficulty have been left on its viscid surface. When the rostellum is touched so quickly that the pollinia are not removed, they are, as I have said, drawn at the moment of explosion a little forward by the movement of the rostellum; but by its subsequent backward movement the pollinia are pushed back into their original position.

From the account here given we may safely infer how the fertilisation of this Orchid is effected. Small insects alight on the broad lower end of the labellum for the sake of the nectar copiously secreted by it; as they lick up the nectar they slowly crawl up its narrowed surface until their heads stand directly under the over-arched crest of the rostellum; as they raise their heads they touch the crest, which explodes, and the pollinia become firmly cemented to them. The insect in flying away withdraws the pollinia, carries them to another flower, and leaves masses of the friable pollen on its viscid stigma.

H

In order to witness what I felt sure would take place, I watched a group of plants on two or three occasions for an hour; each day I saw numerous specimens of two small Hymenopterous insects, namely, a Hæmiteles and a Cryptus, flying about the plants and licking up the nectar; most of the flowers, which were visited over and over again, had already had their pollinia removed, but at last I saw both these insect-species crawl into younger flowers, and suddenly retreat with a pair of bright yellow pollinia sticking to their foreheads; I caught them, and found the point of attachment was to the inner edge of the eye; on the other eye of one specimen there was a ball of the hardened viscid matter, showing that it had previously removed another pair of pollinia, and had subsequently in all probability left them on the stigma of one of the flowers. As I caught these insects, I did not witness the act of fertilisation; but C. K. Sprengel actually saw a Hymenopterous insect leave its pollen-mass on the stigma. My son watched another bed of this Orchid at some miles' distance, and brought me home the same Hymenopterous insects with attached pollinia, and he saw Diptera also visiting the flowers. He was struck with the number of spider-webs spread over these plants, as if the spiders were aware how attractive the Listera was to insects, and how necessary they were to its fertilisation.

To show how delicate a touch suffices to cause the rostellum to explode, I may mention that I found an extremely minute Hymenopterous insect vainly struggling with its whole head buried in the hardened viscid matter, thus cemented to the crest of the rostellum and to the tips of pollinia ; the insect was not so large as one of the pollinia, and after causing the explosion it had not force to remove them, and was thus punished for attempting a work beyond its strength, and perished miserably.

In Spiranthes the young flowers, which have their pollinia in the best state for removal, cannot possibly be fertilised ; they must remain in a virgin condition until they are a little older and the labellum has moved from the column. Here the same thing apparently occurs, for the stigmas of the older flowers were more viscid than those of the younger flowers. These latter have their pollinia quite ready for removal; but immediately after the explosion the rostellum, as we have seen, curls forwards and downwards, thus protecting the stigma for a time, until the rostellum slowly becomes quite straight, when the stigma in its more mature condition is left freely exposed, and is ready for fertilisation.

I was curious to ascertain whether the rostellum would ultimately explode, if never touched; but I have found it difficult to ascertain this fact, as the flowers are so attractive to insects, and the touch of such minute insects suffices to cause the explosion, that it was scarcely possible to exclude them.

I have covered up plants several times, and left them till long after all uncovered plants had set their pods; and without going into unnecessary details, I may positively state that the rostella in several flowers had not exploded, though the stigma was withered, the pollen quite mouldy and incapable of removal. Some few of the very old flowers, however, when roughly touched, were still capable of feeble explosion. Other flowers under the nets had exploded, and the tips of their pollinia were fixed to the crest of the rostellum; but whether these had been touched by some excessively minute insect, or had exploded spontaneously, it is impossible to say. In the case of these latter flowers, it is of some importance to note that not a grain of pollen had got on their stigmas (and I looked carefully), and their ovaria had not swollen. These several facts clearly show that the removal of the pollinia by insect-agency is absolutely necessary to the fertilisation of this species.

That insects do their work effectually, the following cases show :—A rather young spike, with many buds in the upper part, had the pollinia unremoved from the seven upper flowers, but they had been completely removed from the ten lower flowers : there was pollen on the stigmas of six of these ten lower flowers. In two spikes taken together, the twenty-seven lower flowers had their pollinia removed, with pollen on the stigmas of all; these were succeeded by five open flowers with pollinia not removed, and with no pollen on their stigmas ; and these were succeeded by eighteen buds. Lastly, in an older spike with forty-four flowers, all fully expanded, the pollinia had been removed from every one; and on all the stigmas that I examined there was pollen, generally in large quantity.

It will perhaps be worth while to recapitulate the several special adaptations for the fertilisation of this Orchid. The anther-cells open early, leaving the pollen-masses quite loose, with their tips resting on the concave crest of the rostellum. The rostellum then slowly curves over the stigmatic surface, so that its explosive crest stands at a little distance from the anther; and this is very necessary, otherwise the anther would be caught by the viscid matter, and the pollen for ever locked up. This curvature of the rostellum over the stigma and base of the labellum is excellently well adapted to favour an insect striking the crest when it raises its head, after having crawled up the labellum, and licked up the last drop of nectar at its base. The labellum, as C. K. Sprengel has remarked, becomes narrower where it joins the column beneath the rostellum, so that there is no risk of the insect going too much to either side. The crest of the rostellum is so exquisitely sensitive, that a touch from a most minute insect causes it to rupture at two points, and instantaneously two drops of viscid fluid are expelled, which coalesce. This viscid fluid sets hard in so wonderfully rapid a manner that it rarely fails to cement the tips of the pollinia, nicely laid on the crest of the rostellum, to the insect's forehead. As soon as the rostellum has exploded it suddenly curves downwards till it projects at right angles over the stigma, protecting it in its early state from impregnation, in the same manner as the stigma of Spiranthes is protected by the labellum clasping the column. But as in Spiranthes the labellum after a time moves from the column, leaving a free passage for the introduction of the pollinia, so here the rostellum moves back, and not only recovers its former arched position, but stands upright, leaving the stigmatic surface, now become more viscid, perfectly free for pollen to be left on it. The pollen-masses, when once cemented to an insect's forehead, will generally remain firmly attached to it until the viscid stigma of a mature flower removes these encumbrances from the insect, by rupturing the weak elastic threads by which the grains are tied together—receiving at the same time the benefit of fertilisation.

Listera cordata.—Professor Dickie of Aberdeen was so kind as to send me two sets of specimens, but I applied rather too late in the season. The structure is essentially the same as in the last species, and the loculi of the rostellum were very plain. In the middle, on the crest of the rostellum, two or three little hairy points project; whether these have any functional importance I know not. The labellum has two basal lobes (of which vestiges may be seen in L. ovata) which curve up on each side, and would compel an insect to approach the rostellum straight in front. Two flowers had been touched either during the journey, or too quickly by some insect, and had exploded; and their pollinia in consequence were firmly cemented to the crest of the rostellum; but in most of the spikes all the pollen-masses had been removed by insects.

Neottia nidus-avis.—I made numerous ob-

servations on this, the Bird's-nest Orchis,* but
they are not worth giving, as the action and
structure of every part is almost identically
the same as in Listera ovata. The labellum
secretes plenty of nectar, which I mention
merely as a caution, because during one cold
and wet season I looked several times and
could not see a drop, and was perplexed at
the apparent absence of any attraction for
insects; nevertheless, had I looked more per-
severingly, I should probably have found
nectar.

Whether the rostellum ultimately explodes,
if not touched, I could not ascertain; that it
long remains unexploded, though ready to
act, is certain; but I found in 1860 so many
flowers exploded, with a bead of the purplish
hardened cement attached to the crest of the
rostellum and to the unremoved pollinia, that
I suspect that it does explode, after a time,
spontaneously, without the excitement of a
touch. In one large spike, every flower had

* This unnatural sickly-looking plant has generally been supposed to be parasitic on the roots of the trees under the shade of
which it lives; but, according to Irmisch ('Beitrage zur Biologie
und Morphologie der Orchideen,' 1853, s. 25), this certainly is
not the case.

been visited by insects, and all the pollinia
had been removed. Another unusually fine
spike from South Kent, sent me by Mr.
Oxenden, had borne forty-one flowers, and
produced twenty-seven large seed-capsules,
besides some smaller ones.

The pollen resembles that of Listera, in
consisting of grains (formed of four granules)
tied together by a few weak threads; it
differs in being much more incoherent, so that
after a few days it swells and overhangs the
sides and summit of the rostellum: hence,
if the rostellum of a rather old flower be
touched, and an explosion be caused, the pollinia are not so neatly caught by their tips
as in Listera; consequently a good deal of
the friable pollen is often left behind in the
anther-cells and is apparently wasted. Portions of this fall on the corolla; and as the
pollen in this state readily adheres to any
object, it is not improbable that insects
crawling about would thus get dusted, and
leave some on the viscid stigma, without
having touched the rostellum and caused it
to explode. Certainly if the labellum were
more upturned, so that insects were forced to

brush against the anther and column, they would get smeared with the pollen as soon as it had become friable, and might thus effectually fertilise the flower.

This observation interested me, because, when previously examining Cephalanthera, with its aborted rostellum, its upturned labellum, and its friable pollen, I had speculated how a transition, with each gradation useful to the plant, could possibly have been effected from the state of the pollen and flower in the allied Epipactis, with its pollinia attached to a well-developed rostellum, to the present condition of Cephalanthera. Neottia nidus-avis shows to a certain extent how such a transition might have been effected. This Orchid is at present mainly fertilised by means of the explosive rostellum, which acts effectually only as long as the pollen remains in mass; but unless we suppose that the circumstance of the pollen soon becoming friable is a mere injury to the plant, we may believe that the pollen in this condition is sometimes transported to the stigma by its adhesion to the hairy bodies of insects. If this be so, we can see that by a slight change in the form of the

flower, and by the pollen becoming friable at
a still earlier age, this means of fertilisation
might be rendered more and more effectual,
and the explosive rostellum less and less
useful. Ultimately the rostellum would become a superfluity; and then, on the great
principle of the economy of organisation, rendered so necessary by the struggle for life, by
which every part of every being tends to be
saved, the rostellum would be absorbed or
aborted. In this case we should have the
production of a new Orchid in the condition
of Cephalanthera, as far as its means of
fertilisation are concerned, but in general
structure still closely allied to Neottia and
Listera.

CHAPTER V.

Cattleya, simple manner of fertilisation — Masdevallia, curious closed flower — Dendrobium, contrivance for self-fertilisation — Vandeæ, diversified structure of the pollinia; importance of the elasticity of the pedicel; its power of movement — Elasticity and strength of the caudicle — Calanthe with lateral stigmas, manner of fertilisation — Angræcum sesquipedale, wonderful length of nectary — Acropera, perplexing case, a male Orchid.

HAVING examined the means of fertilisation in so many British Orchids, belonging to fourteen genera, I was anxious to ascertain whether the exotic forms, belonging to quite distinct Tribes, equally required insect-agency. I especially wished to ascertain whether the rule holds good that each flower is necessarily fertilised by pollen brought from a distinct flower; and in a secondary degree I was curious to know whether the pollinia underwent those curious movements of depression by which they are placed, after transportal by insects, in the proper position to strike the stigmatic surface.

By the kindness of many friends and strangers I have been enabled to examine fresh flowers of several species, belonging

to forty-three exotic genera, well dispersed
through the sub-families of the vast Orchidean
series.* It is not my intention to describe the
means of fertilisation in all these genera, but
merely to pick out a few curious cases, and
other cases which illustrate the foregoing de-
scriptions. The diversity of the contrivances,
almost all adapted to favour the intercrossing
of distinct flowers, seems to be exhaustless.

* I am particularly indebted to Dr. Hooker, who on every
occasion has given me his invaluable advice, and has never become
weary of sending me specimens from the Royal Gardens at Kew.
Mr. James Veitch, jun., has generously given me many beautiful
Orchids, some of which were of especial service. Mr. R. Parker
also sent me an extremely valuable series of forms. Lady Dorothy
Nevill most kindly placed her magnificent collection of Orchids
at my disposal. Mr. Rucker of West Hill, Wandsworth, sent me
repeatedly large spikes of Catasetum, a Mormodes of extreme
value to me, and some Dendrobiums. Mr. Rodgers of Sevenoaks
has given me interesting information. Mr. Bateman, so well
known for his magnificent work on Orchids, sent me a number
of interesting forms, including the wonderful Angræcum sesquipedale.

I am greatly indebted to Mr. Turnbull of Down for allowing
me the free use of his hot-houses, and for giving me some interesting Orchids; and to his gardener, Mr. Horwood, for his aid in
some of my observations.

Professor Oliver has kindly aided me with his large stores of
knowledge, and has called my attention to several papers. Lastly,
Dr. Lindley has sent me fresh and dried specimens, and has in
the kindest manner helped me in various ways.

To all these gentlemen I can only express my cordial thanks
for their unwearied and generous kindness.

EPIDENDREÆ.

In the arrangement given by Lindley in that invaluable work the Vegetable Kingdom, we first meet with the two great tribes of the Malaxeæ and Epidendreæ. They are characterised by the pollen-grains cohering into large waxy masses, not congenitally attached to the rostellum. In the Malaxeæ (of which one British form has been described) the pollinia properly have no caudicle ; whereas, in the Epidendreæ (of which tribe no British representative exists), they have an unattached caudicle.

For my purpose these two tribes might have been run together ; and as the pollinia of some of the Malaxeæ have a minute but perfectly efficient caudicle, the division between them, in this their leading character, blends away. But this is a misfortune which every naturalist encounters in attempting to classify a largely - developed or so - called natural group, in which, relatively to other groups, there has been little extinction. In order that the naturalist may be enabled to give precise and clear definitions of his

divisions, whole ranks of intermediate or
gradational forms must have been utterly
swept away : if here and there a member
of the intermediate ranks has escaped an-
nihilation, it puts an effectual bar to any
absolutely distinct definition.

I will begin with the genus Cattleya, of
which I have seen several species, and
which is fertilised in a very simple manner,
different from that in any British Orchid.
The rostellum (r, Fig. A, B) is a broad,
tongue-shaped projection, which arches slightly
over the stigma : the upper surface is formed
of smooth membrane ; the lower surface and
the central portions (originally a mass of
cells) consist of a very thick layer of viscid
matter. This viscid mass is hardly separated
from the viscid matter thickly coating the
stigmatic surface which lies close beneath the
rostellum. The generally projecting upper
lip of the anther rests on, and opens close
over, the base of the upper membranous
surface of the tongue-shaped rostellum. The
anther is kept closed by a sort of spring at
the back, at its point of attachment to the
top of the column. The pollinia consist of

Fig. XIX.

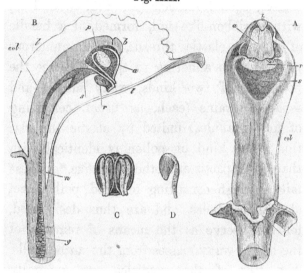

CATTLEYA.

a. anther.	*s*. stigma.
b. spring at the top of the column.	*col*. column.
	l. labellum.
p. pollen-masses.	*n*. nectary.
r. rostellum.	*g*. ovarium, or germen.

A. Front view of column, with all the sepals and petals removed.
B. Section and lateral view of the flower, with all the sepals and petals removed, except the bisected labellum shown only in outline.
C. Anther viewed on the under side, showing the four caudicles with the four pollen-masses beneath.
D. A single pollinium, viewed laterally, showing the pollen-mass and caudicle.

four (or eight in Cattleya crispa) waxy
masses, each furnished (see Fig. C and D)
with a ribbon-like tail, formed of a bundle
of highly elastic threads, with numerous
pollen-grains adhering to them. Hence the
pollen is of two kinds, waxy masses and
separate grains (each, as usual, consisting
of four granules) united by elastic threads:
this latter kind of pollen is identical with
that of Epipactis and other Neotteæ.* These
tails, though consisting of good pollen, act
also as caudicles, and are thus designated,
for they serve as the means of removal of
the larger waxy masses from the anther-cells.
The tips of the caudicles are generally
reflexed, and in the mature flower protrude
a little way out of the anther-case (see
Fig. A), and lie on the base of the upper
membranous lip of the rostellum. The
labellum enfolds the column, making the
flower tubular, and its lower part is produced into a nectary, which penetrates the
ovarium.

---

* The pollen-masses of Bletia are admirably represented on a
gigantic scale in Bauer's drawings, published by Lindley in his
'Illustrations.'

Now for the action of these parts. If any body of size proportional to that of the tubular flower be forced into it — a dead humble-bee acted best — the tongue-shaped rostellum is depressed, and the object often gets slightly smeared with viscid matter; but in withdrawing the object, the tongue-formed rostellum is upturned, and a surprising quantity of viscid matter is forced over its edges and sides, and at the same time into the lip of the anther, which is slightly raised by the upturning of the rostellum. Thus the protruding tips of the caudicles are instantly glued to the retreating body, and the pollinia are withdrawn. This hardly ever failed to occur in my repeated trials. A living bee or other large insect alighting on the fringed edge of the labellum, and scrambling into the flower, would depress the labellum and would be less likely to disturb the rostellum, until it had sucked the nectar and began to retreat. When a dead bee, with the four waxy balls of pollen dangling by their caudicles from its back, is forced into another flower, some or all of these balls are surely caught by the broad, shallow, and very viscid stigmatic surface, which likewise tears off the grains of pollen from the threads of the caudicles.

That living humble-bees can thus remove the pollinia is certain. Sir W. C. Trevelyan sent to Mr. Smith of the British Museum (who forwarded to me) a Bombus hortorum —caught in his hot-house, where a Cattleya was in flower—with its whole back, between the wings, smeared with dried viscid matter, and with the four pollinia attached to it by their caudicles, and ready to be caught by the stigma of another flower if the bee had entered one.

The caudicles of the pollinia being free, and the viscid matter from the rostellum not coming into contact with them without mechanical aid, as well as the general manner of fertilisation, are the same in those species which I have seen of Lælia, Leptotes, Sophronitis, Barkeria, Phaius, Evelyna, Bletia, and Cœlogyne. In Cœlogyne cristata the upper lip of the rostellum is much elongated. In Evelyna caravata eight balls of waxy pollen are all united to one caudicle. In the Barkeria the labellum, instead of enfolding the column, is pressed against it, and this would even more effectually compel insects to brush against the rostellum. In Epidendrum we have a slight difference; for the upper surface of the rostellum, instead of permanently remaining membranous, as in the above-named genera, is so tender that by a touch it breaks up, together with the whole lower surface, into a mass of viscid matter. In this case the whole of the rostellum, with the adherent pollinia, is removed by insects when they retreat from the flower. I observed in E. glaucum that viscid matter exuded from the upper surface of the rostellum when touched, as I have seen in Epipactis; in fact it is difficult to say, in these cases, whether the upper surface of the rostellum should be called membrane or viscid matter.

In Epidendrum floribundum there is a rather greater difference : the anterior horns of the clinandrum (*i. e.* the cup at the summit of the column in which the pollinia lie) approach each other so closely as to adhere to the two sides of the rostellum, which consequently lies in a nick, with the pollinia over

it: and as, in this species, the upper surface
of the rostellum resolves itself into viscid
matter, the pollinia become glued to it without
any mechanical aid. The pollinia, though
thus attached, cannot, of course, be removed
out of their anther - cells without the aid
of insects. In this species it seems possible
(though, from the position of parts, not probable) that an insect might drag the pollinia
on to its own stigma. In all the other species
of Epidendrum which I have examined, and
in all the above-mentioned genera, as the
pollinia lie unattached above the rostellum,
it is evident that the viscid matter has to
be forced upwards into the lip of the anther
by a retreating insect, which would thus necessarily carry the pollinia from one flower to
the stigma of another flower.

MALAXEÆ.

Turning now to the Malaxeæ: in Pleuro-
thallis prolifera and ligulata (?) the pollinia
have a minute caudicle, and mechanical aid
is requisite to force the viscid matter from
the under side of the rostellum into the
anther, thus to catch the caudicles and

remove the pollinia. On the other hand, in our British Malaxis and in the Microstylis Rhedii from India, the upper surface of the minute tongue-shaped rostellum becomes viscid and adheres to the pollinia without mechanical aid. In these two genera we have the curious case of the lower and flat surface of the rostellum being coated with a thin layer of viscid matter, apparently for the purpose of securing the pollinia in a proper position when brought by insects, so as to enter or remain in the slit-like stigma. In Stelis racemiflora the pollinia had also apparently (for the flowers were not in a good condition) become spontaneously attached to the rostellum; and I mention this latter flower chiefly because some insect in the hot-house at Kew had removed most of the pollinia, and had left some of them adhering to the lateral stigmas. These curious little flowers are widely expanded and much exposed; but after a time the three sepals close with perfect exactness and shut up the flower, so that it is scarcely possible to distinguish an old flower from a bud: yet, to my surprise, I found that the closed flowers opened under water.

The allied Masdevallia fenestrata is an
extraordinary flower; for the three sepals,
instead of closing, as in the Stelis after the
flower has remained for a time expanded,
always cohere together and never open.
Two minute, lateral, oval windows (hence
the name fenestrata), seated high up the
flower and opposite each other, afford the
only entrance into the flower; but the presence of these two minute windows (Fig.
XX.) shows how necessary it is that insects

Fig. XX.

Masdevallia fenestrata.
The window on the near side is
shown darkly shaded.
n. nectary.

should have access in this
case as with all other Orchids. How insects per-
form the act of fertilisation
I have failed to understand.
At the bottom of the roomy
and dark chamber formed
by the closed sepals, the
minute column is placed,
in front of which the fur-
rowed labellum stands, with a highly flexible
hinge, and on each side the two upper petals;
a little tube being thus formed. Hence, when
a minute insect enters, or a larger insect inserts
its proboscis through either window, it has by

touch to find the inner tube in order to reach the curious nectary at its base. Within this little tube, formed by the column, labellum, and lateral petals, a very broad and hinged rostellum projects at right angles, the under surface of which is viscid; the minute caudicles of the pollinia, projecting out of the anther-case, rest on the base of the upper membranous surface of the rostellum. The stigmatic cavity is deep. After cutting away the sepals I vainly endeavoured, by pushing a bristle into the tubular flower, to remove the pollinia. The whole structure of the flower seemed carefully intended to prevent the withdrawal of the pollinia, as well as their subsequent insertion into the stigmatic chamber! Some new and curious contrivance has here to be made out.

Of Bolbophyllum I examined the curious little flowers of four species, which I will not attempt fully to describe. In B. cupreum and cocoinum, the upper and lower surfaces of the rostellum resolve themselves into viscid matter, which has to be forced upwards by insects into the anther, so as to secure the pollinia. I effected this easily by passing a needle

down and withdrawing it from the flower,
which is rendered tubular by the position of
the labellum. In B. rhizophoræ the anther-
case moves back, when the flower is mature,
and leaves the two pollen-masses fully exposed
and adhering spontaneously to the upper surface of the rostellum. The two pollen-masses
adhere together by viscid matter, and, judging
from the action of a bristle, are always removed together. The stigmatic chamber is
very deep, and its orifice is oval, and is exactly
fitted by *one* of the two pollen-masses. After
the flower has remained open for some time,
the sides of the oval orifice of the stigmatic
chamber close in and shut it completely,—a
fact which I have observed in no other Orchid,
and which, I presume, is here related to the
much exposed condition of the whole flower.
When the two pollinia were attached to a
needle or bristle, and were forced against the
stigmatic chamber, one of the two masses
glided into the small orifice more readily than
could have been anticipated. Nevertheless it
is evident that insects must place themselves
on successive visits in precisely the same position, so as first to remove the two pollinia,

and then force one of them into the stigmatic orifice. The two upper filiform petals would serve as guides to the insect; but the labellum, instead of making the flower tubular, hangs just like a tongue out of a widely open mouth.

The labellum in all the species which I have seen, more especially in B. rhizophoræ, is remarkable by being joined to the base of the column by a very narrow, thin, white strap, which is highly elastic and flexible; it is even highly elastic when stretched, like an india-rubber band. When the flowers of this latter species were blown by a breath of wind the tongue-like labellums all waggled about in a very odd manner. In some species not seen by me, as in B. barbigerum, the labellum is furnished with a beard of fine hairs, and these cause the labellum to be in almost constant motion from every breath of air. What the use can be of this extreme flexibility and liability to movement in the labellum, I cannot conjecture, unless it be to attract the notice of insects to their dull-coloured, small, and inconspicuous flowers, in the same manner as the bright colours and strong odours of many other Orchids apparently serve to attract insects.

Of the many singular properties of Orchids, the irritability of the labellum in several distantly-allied forms is highly remarkable. When touched, it is described as quickly moving. This is the case with some of the species of Bolbophyllum, but I could not detect any irritability in the species which I examined; nor have I, much to my regret, seen any Orchid with an irritable labellum. The Australian genus Calæna is endowed with this property in a most remarkable degree; for when an insect alights on its labellum it suddenly shuts up against the column, and encloses its prey as it were in a box. Dr. Hooker * believes that this action aids in some manner in the fertilisation of the plant.

The last genus of the Malaxeæ which I will mention is Dendrobium, of which one at least of the species, namely D. chrysanthum, is interesting, from being apparently contrived to effect its own fertilisation, if an insect, when visiting the flower, should accidentally fail to remove the pollen-masses. The rostellum has an upper and a small lower surface composed of membrane; and between these a thick

* 'Flora of Tasmania,' vol. ii. p. 17, under Calæna.

Fig. XXI.

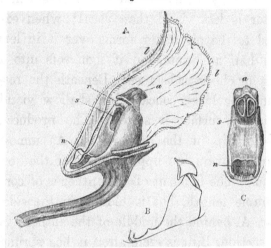

DENDROBIUM CHRYSANTHUM.

a. anther.	*l.* labellum.
r. rostellum.	*n.* nectary.
s. stigma.	

A. Lateral view of flower, with the anther in its proper position,
    before the ejection of the pollinia. All the sepals and
    petals are removed except the labellum, which is longi-
    tudinally bisected.

B. Outline of column, viewed laterally, after the anther has
    ejected the pollinia.

C. Front view of column, showing the empty cells of the anther,
    after it has ejected its pollinia. The anther is represented
    hanging too low, and covering more of the stigma than it
    really does.

mass of milky-white matter is included, which can be very easily forced out. This white matter is less viscid than usual; when exposed to the air a film forms over it in less than half a minute, and it soon sets into a waxy or cheesy substance. Beneath the rostellum the large concave but shallow viscid stigmatic surface is seated. The produced anterior lip of the anther (see A) almost entirely covers the upper surface of the rostellum. The filament of the anther is of considerable length, but is hidden in the side-view, A, behind the middle of the anther; in the section, B, it is seen, after it has sprung forward : it is elastic, and presses the anther firmly down on the inclined surface of the clinandrum (see section B) which lies behind the rostellum. When the flower is expanded the two pollinia, united into a single mass, lie quite loose on the clinandrum and under the anther-case. The labellum embraces the column, leaving a tubular passage in its front; the middle portion (as may be seen in its section, in Fig. A) is thickened ; the thickened portion extends up as far as the top of the stigma. The lowest part of the labellum is developed into a saucer-like nectary, which secretes honey.

If an insect forced its way into one of these flowers, the labellum, which is elastic, would yield, and the projecting lip of the anther would protect the rostellum from being disturbed ; but when the insect retreats, the lip of the anther will be lifted up, and the viscid matter from the rostellum will be forced into the anther, gluing the pollen-mass to the insect, which will thus transport it to another flower. I easily imitated this action ; but as the pollen-masses have no caudicle, and lie rather far back within the clinandrum under the anther, and as the matter from the rostellum is not highly viscid, they were sometimes left behind and not caught.

Owing to the inclination of the base of the clinandrum, and owing to the length and elasticity of the filament, when the anther was lifted up it was always shot over the rostellum, and remained hanging there, with its lower empty surface (Fig. C) suspended over the summit of the stigma. The filament now stretches across the space (see Fig. B) which was originally covered by the anther. Several times, having cut off all the petals and labellum, and laid the flower under the microscope, I raised with a needle the lip of the anther, without disturbing the rostellum, and saw the anther assume, with a spring, the position represented sideways in Fig. B, and frontways in Fig. C. By this springing action the anther scoops the pollen-mass out of the concave clinandrum, and pitches it up in the air, with exactly the right force so as to fall down on the middle of the viscid stigma, where it sticks.

Under nature, however, the action cannot be as thus described; for the labellum hangs downwards; and to understand what follows, the drawing should be placed nearly upside down, in an almost reversed position. If an insect failed to remove the pollinium by means of the viscid matter from the rostellum, the pollinium would first be jerked downwards on to the protuberant surface of the labellum, placed immediately beneath the stigma. But it must be remembered that the labellum is elastic, and that at the same instant that the insect, in the act of leaving the flower, lifted up the lip of the anther, and so caused the pollen-mass to be shot out, the labellum would rebound back, and striking the pollen-mass would pitch it upwards, so as to hit the sticky stigma. Twice I succeeded in effecting this, with the flower held in its natural position, by imitating the retreat of an insect; and on opening the flower I found the pollen-mass glued to the stigma.

This view of the use of the elastic filament, seeing how complicated the action must be, may appear fanciful; but we have seen so many and such curious adaptations, that I cannot believe the strong elasticity of the filament and the thickening of the middle of the labellum to be useless points of structure. If the action be as I have described, it might be an advantage to the plant that its single large pollen-mass should not be wasted, if it failed to adhere to an insect by means of the viscid matter from the rostellum. This contrivance is not common to all the species of the genus; for in neither D. bigibbum nor D. formosum was the filament of the anther elastic, nor was the middle line of the label-lum thickened. In D. tortile the filament was elastic; but as I saw only one flower,

before I had made out the structure of D.
chrysanthum, I cannot say how it acts.

VANDEÆ.

We now come to Lindley's immense Tribe
of the Vandeæ, which includes many of the
most extraordinary productions of our hot-
houses, but has no British representative.
I have examined twenty-four genera. The
pollen consists of waxy masses, as in the two
last Tribes, and each ball of pollen is furnished with a caudicle, which becomes, at
an early period of growth, united to the
rostellum. The caudicle is seldom attached
directly to the viscid disc, as in the Ophreæ,
but to the upper and posterior surface of the
rostellum; and this part, together with the
disc, is removed by insects. The imaginary
diagram (Fig. XXII.), with the parts separated, will best explain the type-structure
of the Vandeæ. The middle organ (2) is
the dorsal or posterior pistil of the three
pistils always present in Orchids; its upper
part is modified into and forms the rostellum, and is curved over the stigma. The
stigma consists of two stigmas belonging to

Fig. XXII.

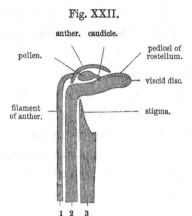

DIAGRAM, illustrative of the structure of the column in the
VANDEÆ.

(1.) The filament, bearing the anther with its pollen-masses:
the anther is represented after it has opened along its
whole under surface, which consequently does not
appear in the section here given.
(2.) The upper pistil modified into the rostellum.
(3.) The two lower confluent pistils, bearing the two confluent
stigmas.

the two other confluent (3) pistils. On the
left hand we have the filament (1) bearing
the anther. The anther opens at an early
period, and the tips of the two caudicles
protrude through a small slit (only one
caudicle and one pollen-mass is represented
in the diagram) in a not fully-hardened
condition, and adhere to the back of the

rostellum. The surface of the rostellum is generally hollowed out for the reception of the pollen-masses; it is represented as smooth in the diagram, but is really often furnished with crests or knobs for the attachment of the two caudicles. The anther afterwards opens more widely on its under surface, and leaves the two pollen-masses unattached, excepting by their caudicles to the rostellum.

During this early period of growth, a remarkable change has been going on in the rostellum : either its extremity or its lower surface becomes excessively viscid, and a line of separation, at first appearing as a mere hyaline zone of tissue, gradually is formed, which separates the viscid extremity or disc, as well as the whole upper surface of the rostellum, as far back as the point of attachment of the caudicles. If any object now touches the viscid disc, it, the whole back of the rostellum, the caudicles and pollen-masses, can all be readily removed together. In botanical works the whole structure between the disc (generally called the gland) and the waxy balls of pollen is designated as the caudicle; but as these parts play an

essential part in the fertilisation of the flower,
and as they are fundamentally different in
their origin and in their minute structure,
I shall call the two elastic ropes, which are
developed strictly within the anther-cells, the
caudicles; and the portion of the rostellum to
which the caudicles are attached (see diagram),
and which is not viscid, the pedicel. The
viscid portion of the rostellum I shall call,
as heretofore, the viscid disc. The whole may
be conveniently spoken of as the pollinium.

In the Ophreæ we always have (except in
O. pyramidalis) two separate viscid discs.
In the Vandeæ, with the exception of Angræcum, we have only one disc. The disc
is naked, or is not enclosed in a pouch. In
Habenaria the discs, as we have seen, are
separated from the two caudicles by short
drum-like pedicels, answering to the single
and generally much more largely developed
pedicel in the Vandeæ. In Ophreæ the
caudicles of the pollinia, though elastic, are
rigid, and serve to place the packets of
pollen at the right distance from the insect's
head or proboscis, so as to reach the stigma.
In the Vandeæ this end is gained by the

pedicel of the rostellum. The two caudicles
in the Vandeæ are attached and embedded
within a deep cleft in the pollen-masses,
and until stretched are rarely visible, for the
pollen-masses lie close to the pedicel of the
rostellum. These caudicles answer both in
position and function to the elastic threads,
by which the packets of pollen are tied
together in the Ophreæ, at the point where
they become confluent and where they form
the upper part of the caudicle; for the function of the true caudicle in the Vandeæ is to
break when the masses of pollen, transported
by insects, adhere to the stigmatic surface.

In many Vandeæ the caudicles are easily
ruptured, and the fertilisation of the flower,
as far as this point is concerned, is a simple
affair; but in other cases the strength of the
caudicles and the length to which they can be
stretched before they break is surprising. I
was at first perplexed to understand what
good purpose the great strength of the
caudicles and their capacity of extension
could serve. It is obvious that, when
projecting far out from an insect's head,
whilst flying about (and the insect, in the

case of the larger Orchids, must be of considerable size), the strength of the caudicles
would protect the pollen-masses from being
brushed off and lost. So again, when an
insect transporting a pollinium visits a flower
either too young, with its stigma not yet
sufficiently viscid, or one already impreg-
nated, with its stigma beginning to dry, the
strength of the caudicle would prevent the
pollen-masses from being uselessly removed.
It should be remembered that the pollen-
masses are precious objects, for each flower,
in most of the genera, produces only two ;
and in many cases, judging from the size of
stigma, both pollen-masses would be left on
one stigma, though in other cases the size of
the orifice of the stigma allows the introduction of one pollen-mass alone ; so that, in
this latter case, the pollen from one flower
probably suffices to fertilise two flowers.

Although at the proper period the stigmatic surface is astonishingly viscid in many
cases, as in Phalænopsis and Saccolabium,
yet when, having removed the pollinia
adhering to a rough scalpel by their viscid
discs, I inserted the balls of pollen into

the stigmatic chamber, they did not adhere
to the surface with sufficient force to prevent
their withdrawal. I even left them for some
little time in contact with the viscid surface,
as an insect would do whilst feeding; but
when I pulled the pollinia straight out of
the stigmatic chamber, the caudicles, though
they were stretched to a great length, did
not rupture, nor did the viscid disc separate
from the scalpel; consequently the balls of
pollen were not left on the stigma. It then
occurred to me that an insect in flying away
would not pull the pollinia straight out of
the chamber, but would pull at nearly right
angles to its orifice. When I thus acted, the
stretched caudicles were necessarily dragged
over the margin of the chamber, and the
friction thus caused, together with the
viscidity of the stigmatic surface, generally
ruptured them and left the pollen-masses on
the stigma. Thus, it seems that the great
strength and extensibility of the caudicles,
which, until stretched, lie embedded within
the pollen-masses, serve to protect the
pollen-masses from being wasted, and yet,
by friction being brought into play, allow

them, at the proper time and by means of
insects, to be left adhering to the stigmatic
surface, and the fertilisation of the flower to
be safely effected.

The discs and pedicels of the rostellum in
the Vandeæ present great diversities in
shape, and an apparently exhaustless number
of adaptations. Even in species of the
same genus, as in Oncidium, these parts
differ greatly. I have here given a few
figures (Fig. XXIII.), taken almost at
hazard. The pedicel generally consists (as
far as I have seen) of a piece of thin ribbon-

Fig. XXIII.

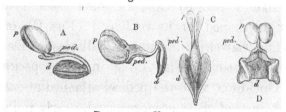

POLLINIA OF VANDEÆ.

d. viscid disc.   *ped.* pedicel.   *p.* pollen-masses.

The caudicles, being embedded within the pollen-masses, are
    not shown.
A. Pollinium of Oncidium grande after partial depression.
B. Pollinium of Brassia maculata (copied from Bauer).
C. Pollinium of Stanhopea saccata after depression.
D. Pollinium of Sarcanthus teretifolius after depression.

shaped membrane (Fig. A), long or short;
but sometimes it is almost (Fig. C) cylindrical
and often of all sorts of shapes. The pedicel
is generally nearly straight, but in Miltonia
Clowesii it is naturally curved; and in some
other cases, as we shall immediately see, it
assumes, after removal, various shapes. The
extensible and elastic caudicles, by which the
pollen-masses are attached to the pedicel,
are here not visible, being embedded in a
cleft or hollow within each pollen-mass. The
disc, which is viscid on the under side, consists of a piece of thin or thick membrane of
the most diversified shapes. In Acropera it
is like a pointed cap; in some cases it is
tongue-shaped, or heart-shaped (Fig. C), or
saddle-shaped, as in some Maxillarias, or like
a thick cushion (Fig. A), as in many species
of Oncidium, with the pedicel attached at one
end, instead of, as is more usual, nearly to its
centre. In Angræcum distichum and sesquipedale the rostellum is notched, and two
separate, thin, membranous discs can be removed, each carrying by a short pedicel its
pollen-mass. In Sarcanthus teretifolius the
disc (Fig. D) is very oddly shaped; and as

the stigmatic chamber is very deep, and likewise curiously shaped, one is tempted to believe that the disc has to be fastened with great precision on to the square projecting head of some insect.

In most cases there is a plain relation between the length of the pedicel and the depth of the stigmatic chamber, into which the pollen-masses have to be inserted; in some few cases, however, in which there is a long pedicel and a shallow stigma, we shall meet with curious compensating actions. After the disc and pedicel have been removed, the shape of the rostellum is altered, being generally only slightly shortened and made thinner: sometimes it becomes notched: in Stanhopea, the entire circumference of the extremity of the rostellum is removed, and a thin, pointed, needle-like process alone is left, which originally ran up its centre.

If we now turn to the former imaginary diagram (Fig. XXII.,.p. 179), and suppose the rectangularly bent rostellum to be thinner and the stigma to lie closer under it, we shall see that, if an insect with the pollinium attached to its head were to fly to another flower and

occupy almost exactly the same position
which it held when the attachment of the
disc was effected, the pollen-masses would
strike the stigma, especially if, from their
weight, they had become in the least degree
depressed. This is all that takes place in
Lycaste Skinnerii, Cymbidium giganteum,
Zygopetalum Mackai, Angræcum eburneum,
Miltonia Clowesii, in a Warrea, and, I believe,
in Galeandra Funkii. But if in our diagram
we suppose the stigma to be seated lower down
at the bottom of a deep cavity, or suppose
the anther to be seated higher up so that the
pedicel of the rostellum sloped upwards, and
under other contingencies not worth detailing,
all of which occur,—in such cases, an insect
with the pollinium attached to its head, on
flying to another flower would not strike the
stigma with the pollen-masses unless some
great change had intervened in their position
after attachment.

This change is effected in many Vandeæ
in the same manner as is so general with
the Ophreæ, namely, by a movement of
depression in the pollinium in the course of
about half a minute after its removal from

the rostellum. I have seen this movement conspicuously displayed, generally causing the pollinium after detachment to rotate through about a quarter of a circle, or 90°, in several species of Oncidium, Odontoglossum, Vanda, Aerides, Sarcanthus, Saccolabium, Acropera, and Maxillaria. In Rodriguezia suaveolens the movement of depression is remarkable from its extreme slowness ; in Eulophia viridis from its small extent. In some, indeed, of the cases specified of the pollinia undergoing no movement, I am not sure that there is not a very slight depression. In the Ophreæ the anther-cells are sometimes seated exteriorly and sometimes interiorly with respect to the stigma ; and we accordingly have outward and inward movements in the pollinia : but in the Vandeæ the anther-cells lie directly over the stigma, and the movement of the pollinium is always directly downwards. In Calanthe, however, the two stigmas are placed exteriorly to the anther-cells, but the pollinia, as we shall see, are made to strike them by a mechanical arrangement of the parts.

In the Ophreæ the seat of contraction,

which causes the act of depression, is in the upper surface of the viscid disc, near the point of attachment of the caudicle: in the Vandeæ the seat is likewise in the upper surface of the viscid disc, but at the point where the pedicel is united to it, and therefore at a considerable distance from the point of attachment of the true caudicles. The contraction and consequent movement is hygrometric in its nature (but the subject, as we shall see in the seventh chapter, is rather obscure), and consequently does not take place until the pollinium is removed from the rostellum, and the point of union of the disc and pedicel has been exposed for a few seconds to the air. If, after the contraction and consequent movement of the pedicel, the whole body be placed into water, the pedicel slowly moves back and resumes the same position with respect to the viscid disc which it held when forming part of the rostellum. When taken out of water, it again undergoes the movement of depression. It is of importance to notice these facts, as we thus get a test by which this movement can be distinguished from certain other movements.

In one species of Maxillaria, viz. in the M. ornithorhyncha, we have a unique case. The pedicel of the rostellum is much elongated, and is entirely covered by the produced front lip of the anther, and is thus kept damp. When removed there is no movement at the junction of the disc and pedicel; but the pedicel itself, at a point rather more than half-way up, quickly bends backwards on itself, in a reversed direction compared with all other cases. When the pedicel is placed in water it resumes its original straight form. If we suppose the long upright neck of a bird to represent the pedicel, the head representing the balls of pollen, then in all ordinary cases the movement is that of a bird picking up food from the ground, but bending only the lower vertebræ close to its body; whereas, in this Maxillaria, the movement is as if the bird threw its head backwards so as nearly to touch its own back, the middle vertebræ of the neck being alone bent. I stated above that, when the pedicel is long and the stigmatic cavity shallow, as in this Maxillaria, we have a compensating action; and here we have one instance. The labellum

has a square projection in front of the stigma,
the passage into the flower being thus contracted; and if the pedicel of the rostellum
did not somehow become shortened, the flower
could hardly be fertilised. After the reversed
movement just described, and the consequent
shortening of the pedicel, the pollinium, when
attached to any small object, can be inserted
into the flower, and the balls of pollen are so
placed as readily to adhere to the stigmatic
surface.

In some cases, besides these hygrometric
movements, elasticity comes into play. In
Aerides odorata and virens, and in Oncidium
(roseum?), the pedicel of the rostellum is
fastened down in a straight line, by the disc
at one extremity and by the anther at the
other; it has, however, a strong natural elastic
tendency to spring up at right angles to the
disc. Consequently, when the pollinium is
removed by its viscid disc sticking to any
object, the pedicel instantly springs up and
stands at nearly right angles to its former
position, with the pollen-masses carried aloft.
This has been noticed by other observers;
and I agree with them that the object gained

is to free the pollen-masses from their anther-cells. After this upward elastic spring, the downward hygrometric movement immediately commences, which, oddly enough, carries the pedicel back again into almost the same position, relatively to the disc, which it held whilst adhering to the rostellum. The end of the pedicel, however, which in Aerides carries the pollen-mass by short dangling caudicles, remains curved a little upwards ; and this seems well adapted to drop the pollen-masses into the deep stigmatic cavity over a ledge in front. The difference between the first elastic and the second or reversed hygrometric movement, was well seen (I tried it only in the Oncidium) by placing the pollinium, after both movements had taken place, into water, when the pedicel moved into the position which it had at first acquired by elasticity, which latter position was not in any way affected by the water. When taken out of water the hygrometric movement of depression soon recommenced.

In Rodriguezia secunda there was no slow movement of depression in the pedicel as in

K

the before-mentioned R. suaveolens, but there
was a rapid downward movement, due, apparently, in this one case, to elasticity; for
when the pedicel was put into water it
showed no tendency to recover its original
position, as occurs with all the many other
Orchids which I have examined.

In Phalænopsis grandiflora and amabilis
the stigma is shallow and the pedicel of the
rostellum long. Consequently a compensating
action is requisite, which, differently from the
case of the Maxillaria, is effected by elasticity.
There is no movement of depression; but,
when the pollinium is removed, the straight
pedicel suddenly curls up, thus (·‒⌒‒). The
full-stop on the left hand may represent the
balls of pollen, but the disc on the right
hand must be imagined to be a triangular
piece of membrane. The pedicel does not
straighten itself in water. The end carrying
the balls of pollen after the contraction is a
little raised up, and the pedicel, with one
end raised, and with the middle part up-
wardly bowed, seems well adapted to drop
the pollen-masses, over a ledge in front, into
the deep stigmatic cavity.

In Calanthe Masuca and the hybrid C. Dominii the structure is very different. We here have two oval, pit-like stigmas standing quite laterally on each side of the rostellum (Fig. XXIV.). The viscid disc is oval (Fig. B), and has no pedicel, but eight masses

Fig. XXIV.

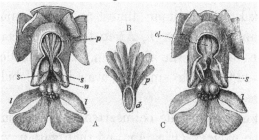

CALANTHE MASUCA.

p. pollen-masses.	*d.* viscid disc.
s s. the two stigmas.	*cl.* clinandrum with pollen-
n. mouth of nectary.	masses removed.
l. labellum.	

A. Flower viewed from above, with the anther-case removed, showing the eight pollen-masses in their proper position within the clinandrum. All the sepals and petals have been cut away except the labellum.

B. Pollen masses attached to the viscid disc, seen from the under side.

C. Flower in same position as in A, but with the disc and pollen-masses removed, showing the now divided rostellum and the empty clinandrum in which the pollen-masses lay. Within the left-hand stigma two pollen-masses may be seen adhering to its viscid surface.

of pollen are attached to it by very short
and easily ruptured caudicles. These pollen-
masses radiate from the disc like the spokes
of a fan. The rostellum is broad and its sides
slope on each side towards the lateral pit-
like stigmas. When the disc is removed
the rostellum is seen (Fig. C) to be deeply
divided in the middle. The labellum is
united to the column almost up to its summit,
leaving a passage (n, A) to the long nectary
close beneath the rostellum. The labellum
is studded with singular, wart-like, globular
excrescences.

If a thick needle be inserted into the mouth
of the nectary (Fig. A), and then withdrawn,
the viscid disc will be withdrawn, bearing
with it the elegant fan of radiating pollen-
masses. These undergo no change in position. But if the needle be now inserted into
the nectary of another flower the ends of the
pollen-masses necessarily hit the upper and
laterally sloping sides of the rostellum, and,
glancing off both ways, they strike down into
the two lateral pit-like stigmas. The thin
caudicles being easily ruptured, the pollen-
masses are left adhering like little darts (see

left-hand stigma in Fig. C) to the viscid surface
of both stigmas, and the fertilisation of the
flower is completed in a simple manner pleas-
ing to behold.

I should have stated that a narrow transverse rim of stigmatic tissue, running beneath the rostellum, connects the two lateral
stigmas; and it is probable that some of the
middle pollen-masses may be inserted through
the notch in the rostellum or beneath its
surface. I am more inclined to this opinion
from having found in the elegant Calanthe
vestita that the rostellum extends so widely
over the two lateral stigmas, that apparently
all the pollen-masses would have to be inserted
beneath its surface.

I fear that the reader will be wearied, but
I must say a few words on the Angræcum
sesquipedale, of which the large six-rayed
flowers, like stars formed of snow-white wax,
have excited the admiration of travellers in
Madagascar. A whip-like green nectary of
astonishing length hangs down beneath the
labellum. In several flowers sent me by
Mr. Bateman I found the nectaries eleven
and a half inches long, with only the lower

inch and a half filled with very sweet nectar.
What can be the use, it may be asked, of a
nectary of such disproportional length? We
shall, I think, see that the fertilisation of the
plant depends on this length and on nectar
being contained only within the lower and
attenuated extremity. It is, however, surprising that any insect should be able to
reach the nectar : our English sphinxes have
probosces as long as their bodies; but in
Madagascar there must be moths with probosces capable of extension to a length of
between ten and eleven inches !

The rostellum is broad and foliaceous, and
arches rectangularly over the stigma and
over the orifice of the nectary : it is deeply
cleft, with the cleft enlarged or widened at
the end. Hence the rostellum pretty closely
resembles (see Fig. XXIV., C) that of
Calanthe after the disc has been removed.
The under surfaces of both margins of the
cleft near its end are bordered by narrow
strips of viscid membrane, easily removed ;
so that there are two distinct viscid discs.
To the middle of each disc a short membranous pedicel is attached ; and each pedicel

carries at its other end a pollen-mass. Beneath the rostellum a narrow, ledge-like, viscid stigma is seated.

I could not for some time understand how the pollinia of this Orchid were removed, or how it could be fertilised. I passed bristles and needles down the open entrance into the nectary and through the cleft in the rostellum with no result. It then occurred to me that, from the length of the nectary, the flower must be visited by large moths, with a proboscis thick at the base ; and that to drain the last drop of nectar even the largest moth would have to force its proboscis as far down as possible. To effect this, whether or not the moth first inserted its proboscis by the open entrance into the nectary (as is most probable, from the shape of the flower, &c.) or through the cleft in the rostellum, it would ultimately force its proboscis into this cleft, for this is the straightest course, and by slight pressure the whole foliaceous rostellum can be depressed : the distance from the outside of the flower to the extremity of the nectary can be thus shortened by about a quarter of an inch. Hence I took a cylinder, one-tenth of an inch

in diameter, and pushed it down through the
cleft in the rostellum : the margins readily
separated, and were pushed downwards together with the whole rostellum. When I
slowly withdrew the cylinder the rostellum
rose from its elasticity, and the margins of
the cleft were upturned and clasped the
cylinder. Thus the viscid strips of membrane on the under sides of the cleft rostellum
came into contact with the cylinder, and
firmly adhered to it; and the pollen-masses
were withdrawn. By this means alone I
succeeded in each case in withdrawing the
pollinia; and it cannot, I think, be doubted
that a large moth must thus act; namely,
by driving its proboscis up to the very base,
through the cleft of the rostellum, so as to
reach the extremity of the nectary; and then
withdrawing its proboscis with the pollinia
attached to it.

I did not succeed in imitating the fertilisation of the flower so well as I did in
withdrawing the pollinia, but I effected it
twice. As the margins of the cleft rostellum
must be upturned before the discs adhere to
the cylinder, they become, during its with-

drawal, affixed some little way from its actual
base. The two discs did not always adhere
at exactly corresponding points. Now, when
a moth inserts its proboscis, with the pollinia
affixed to it near the base, into the mouth of
the nectary, the pollen-masses will probably
be first inserted beneath the rostellum; and
during the final exertion, when the moth
pushes its proboscis through the cleft of the
rostellum, the pollen-masses will almost necessarily be placed on the narrow, ledge-like
stigma projecting beneath the rostellum. By
acting thus with the pollinia attached to the
cylinder the pollen-masses were twice torn off
and left glued to the stigmatic surface.

If the Angræcum in its native forests
secretes more nectar than did the vigorous
plants sent me by Mr. Bateman, so that the
nectary becomes filled, small moths might
obtain their share, but they would not benefit
the plant. The pollinia would not be withdrawn until some huge moth, with a wonderfully long proboscis, tried to drain the last
drop. If such great moths were to become
extinct in Madagascar, assuredly the Angræcum would become extinct. On the other

hand, as the nectar, at least in the lower part
of the nectary, is stored safe from depreda-
tion by other insects, the extinction of the
Angræcum would probably be a serious loss
to these moths. We can thus partially
understand how the astonishing length of the
nectary may have been acquired by successive
modifications. As certain moths of Madagascar became larger through natural selection in relation to their general conditions
of life, either in the larval or mature state,
or as the proboscis alone was lengthened
to obtain honey from the Angræcum and
other deep tubular flowers, those individual
plants of the Angræcum which had the
longest nectaries (and the nectary varies much
in length in some Orchids), and which,
consequently, compelled the moths to insert
their probosces up to the very base, would
be best fertilised. These plants would yield
most seed, and the seedlings would generally
inherit longer nectaries ; and so it would be
in successive generations of the plant and
moth. Thus it would appear that there has
been a race in gaining length between the
nectary of the Angræcum and the proboscis

of certain moths; but the Angræcum has triumphed, for it flourishes and abounds in the forests of Madagascar, and still troubles each moth to insert its proboscis as far as possible in order to drain the last drop of nectar.

Finally, another genus, Acropera, must be noticed from an independent reason. Although Dr. Hooker sent me, over and over again, fresh flowers of two species (A. luteola * and Loddigesii), this genus for a long time remained the opprobrium of my work. All the parts seemed determinately contrived that the plant should never be fertilised. I believe I have at last partly solved the mystery; yet the use of some important parts remains quite unintelligible. But I do not suppose that I completely understand the contrivances in any one Orchid; for I find that the more I study our commonest British species, continually new and admirable adaptations become apparent.

* Dr. Lindley informs me that he knows of no species so named; nor is the origin of the name known at Kew. This species or variety differs in little or in nothing from A. Loddigesii, excepting in its uniform yellow colour.

The rostellum of Acropera is thin and
elongated, projecting at right angles to the
column (see diagram, Fig. XXII., p. 179);
the pedicel of the pollinium is of course
equally long and very thin; the disc is
extraordinarily small, and forms a little cap,
viscid within, fitting the extremity of the rostellum. After repeated trials I find that the
disc does not adhere to any object until it is
drawn quite off the tip of the rostellum; and
this can only be well effected by the whole
rostellum being pushed upwards so as to slide
over and against the touching object: when
the small disc is thus removed it adheres
well to the object. The upper sepal forms
a hood enclosing and protecting the column.
The labellum is an extraordinary organ, baffling
description: it is articulated to the base of
the column by a thin strap, so elastic and
flexible that a breath of wind sets it vibrating. It hangs downwards; and this seems
to be of importance, for the plant is pen-
dulous, and to place the labellum in this
position the footstalk (ovarium) of each
flower is curved into a semicircle. The two
upper petals serve as lateral guides leading

into the hood-like upper sepal. But how all these parts concur in leading an insect to push some part of its body into the hood-like upper sepal and then to raise the rostellum, thus brushing off the little sticky disc, I do not in the least understand.

The pollinium, when adhering by its disc to an object, undergoes the common movement of depression; and this seems superfluous, for the stigmatic cavity lies (see diagram, Fig. XXII.) high up at the base of the rectangularly projecting rostellum. But this is a comparatively trifling difficulty; the real difficulty lies in the orifice of the stigmatic chamber being so narrow that the pollen-masses can hardly be forced in. I repeatedly tried, and succeeded only three or four times. Even after allowing the pollen-masses to dry for an hour, and thus to shrink a little, I rarely succeeded in forcing them in. I examined young flowers and almost withered flowers, for I imagined that the mouth of the chamber at different periods of growth might become larger or smaller, for we have seen that the mouth actually closes up in one species of Bolbophyllum ; but the difficulty of insertion always remained the same. Now when we observe, that the viscid disc is extraordinarily small, and consequently its power of attachment not so firm as with Orchids having a large disc, and that the pedicel is long and thin, it would seem almost indispensable that the stigmatic chamber should be unusually large for the easy insertion of the pollinium. Far from this being the case, it is, as just stated, so much contracted that rarely by any force could even one pollen-mass be forced in. Moreover the stigmatic surface, as Dr. Hooker also observed, is singularly little viscid!

I had given up the whole case as inexplicable, when it occurred to me that, although no instance of the separation of the two sexes was known in Orchids, yet that Acropera might be a male plant. I first examined the utriculi of the stigmatic surface from specimens which had been kept in spirits of wine, and I found them empty like little glass cases, but generally with a faint areola or nucleus visible.* Now I have looked at the

---

* See R. Brown in ' Linn. Transact.,' vol. xvi. p. 710, on the nucleus of the stigmatic utriculi; and Bauer's beautiful drawing in Lindley's great work.

utriculi of very many Orchids, and have
hitherto found no exception to the rule that
spirits of wine causes a considerable quantity
of yellowish-brown matter to coagulate within
them. I have taken fresh Orchids and placed
them in spirits, and in twenty-four hours have
found the contents coagulated and the nuclei
much darkened. More numerous observations
would be requisite for much stress to be laid
on this fact, but I must at present infer that
the utriculi of Acropera are in a different condition from those of all other Orchids. The
state of the ovarium offers better evidence.

When a thin transverse slice of the ovarium
is taken and examined under a quite weak
power, small projections are seen on the
three proper ovule-bearing cords or segments,
which at first seem like true ovules. But
when these are more closely examined, they
are seen to consist of sub-branched, quite thin
and transparent fringes of membrane, which
in some specimens exhibited cellular structure
far more plainly than in others. If these
fringes are placentæ, they are more largely
developed than in other Orchids; if they are
ovules (or rather the testæ of ovules) in an

atrophied condition, as I believe to be the
case, they are more firmly fixed to the placentæ than in other cases; they do not exhibit
the proper opening at their free ends, and
no nucleus is visible; nor were any of them
inverted. I examined six ovaria of young
and old flowers of the Acropera, some fresh
and some which had been kept in spirits of
wine, and all the ovule-bearing cords were
nearly in the same condition. I examined,
for comparison, the ovaria of Orchids belonging to nearly all the main Tribes, of young
and old (but not fertilised) flowers, some of
which were fresh, and some kept in spirits,
and invariably the ovules presented a widely
different appearance.

From these several facts—namely, the narrowness of the mouth of the stigmatic chamber,
into which the pollen-masses can hardly be
forced, whereas the length and thinness of
the pedicel of the rostellum, the smallness of
the viscid disc, and the movement of depression, all indicate the necessity of a large
stigmatic cavity seated low down—the slight
viscidity of the stigmatic surface—the empty
condition of the stigmatic utriculi—and espe-

cially the condition of the ovule-bearing cords—lead me to infer that the plant at Kew, from which the many flowers of the Acropera luteola were at different times gathered, is a male plant. From having examined many Orchids grown in hot-houses,* I have no reason to believe that cultivation affects the female organs in the manner described. It is scarcely possible to believe that cultivation could contract the solid edges of the stigmatic chamber. There-fore I see no reason to doubt my conclusion on the male sex of this plant.

What the female or hermaphrodite form of the Acropera luteola may prove to be— whether resembling in most respects the male, or whether it be at present named and masked as some distinct genus—it it impos-sible to say. In Acropera Loddigesii, which closely resembles, in all respects except in colour, A. luteola, I found the same almost insuperable difficulty in inserting the pollen-

* In a spike of the Brazilian Goodyera discolor, sent me by Mr. Bateman, which had all its flowers monstrous and distorted, with the stigmas imperfect, I found the ovules with their nuclei projecting far out of the testæ (exactly as figured by Brongniart in Epipactis in ' Annales des Sciences,' tom. 24, 1831, pl. 9), and apparently well developed.

masses into the stigmatic cavity, but I did not
at that time suspect the masculine nature of
the genus, and did not examine the ovarium.

I have now described, perhaps in too much
detail, a few of the many contrivances by
which the Vandeæ are fertilised. The relative position of the parts—friction, viscidity,
elastic and hygrometric movements, all nicely
related to each other—come into play. But
all these appliances are subordinate to the
action of insects. Without their aid, not a
plant in this tribe, in the twenty-four genera
examined, would set a seed. It is also evident that, in a vast majority of cases, insects
would withdraw the pollinia only when retreating from a flower, and, carrying them
away, would thus effect a union between
two distinct flowers. This fact is conclusively
shown in all those many cases in which the
pollinia undergo a change of position, after
removal from the rostellum, in order to stand
in a proper direction to strike the stigma;
for this could only be effected after the insect
had left one flower, which would serve as the
male, and before it visited a second flower,
which would serve as the female.

CHAPTER VI.

Catasetidæ, the most remarkable of all Orchids — The mechanism by which the pollinia of Catasetum are ejected to a distance, and are transported by insects — Sensitiveness of the horns of the rostellum — Extraordinary difference in the male, female, and hermaphrodite forms of Catasetum tridentatum — Mormodes ignea, curious structure of flower; ejection of its pollinia — Cypripedium, importance of the slipper-like form of the labellum — Secretion of nectar — Advantage derived from insects being delayed in sucking the nectar — Singular excrescences on the labellum, apparently attractive to insects.

I HAVE reserved for separate description one sub-family of the Vandeæ, namely, the Catasetidæ, which may, I think, be considered as the most remarkable of all Orchids.

I will begin with the most complex genus, Catasetum. A brief inspection of the flower shows that here, as with other Orchids, some mechanical aid is requisite to remove the pollen-masses from their receptacles, and to carry them to the stigmatic surface. We shall, moreover, presently see that the three following species of Catasetum are male plants; hence it is certain that their pollen-masses must be transported to female plants,

in order that seed may be produced. The
pollinium is furnished with a viscid disc, in this
genus of huge size ; but the disc, instead of
being placed, as in other Orchids, in a position
likely to touch and adhere to an insect visiting the flower, is turned inwards and lies close
to the upper and back surface of a chamber,
which must be called the stigmatic chamber,
though functionless as a stigma. There is
nothing in this chamber to attract insects;
and even if they did enter it, it is hardly
possible that the disc should adhere to them,
for its viscid surface lies in contact with the
roof of the chamber.

How then does Nature act ? She has
endowed these plants with, what must be
called for want of a better term, sensitiveness,
and with the remarkable power of forcibly
ejecting their pollinia to a distance. Hence,
when certain definite points of the flower are
touched by an insect, the pollinia are shot out
like an arrow which is not barbed, but has a
blunt and excessively adhesive point. The
insect, disturbed by so sharp a blow, or after
having eaten its fill, flies sooner or later to
a female plant, and, whilst standing in the

same position as it did when struck, the pollen-bearing end of the arrow is inserted into the stigmatic cavity, and a mass of pollen is left on its viscid surface. Thus, and thus alone, at least three species of the genus Catasetum are fertilised.

In many Orchids, as in Listera, Spiranthes, Orchis, we have seen that the surface of the rostellum is so far sensitive, that, when touched or when exposed to the vapour of chloroform, it ruptures in certain defined lines. So it is in the tribe of the Catasetidæ, but with this remarkable difference, that in Catasetum the rostellum is prolonged into two curved taper-ing horns, or, as I shall call them, antennæ, which stand over the labellum where insects alight, and the excitement of a touch is conveyed along these antennæ to the mem-brane which has to be ruptured; and when this is effected, the disc of the pollinium is suddenly set free. We have also seen that in several Vandeæ the pedicels of the pollinia are fastened down flat, but are elastic and tend to spring up, so that, as soon as they are freed, they suddenly curl upwards, apparently for the purpose of detaching the pollen-masses

from their anther-cells. In the genus Catasetum, on the other hand, the pedicels are
fastened down in a curved position; and when
freed by the rupture of the attached edges of
the disc, they straighten themselves with such
force, that not only do they drag the balls of
pollen and anther-cells from their places of
attachment, but the whole pollinium is jerked
forward, over and beyond the tips of the so-
called antennæ, to the distance of two or
three feet. Thus, as throughout nature, pre-
existing structures and capacities are utilised
for new purposes.

Catasetum saccatum.*—I will now enter on
details. The general appearance of this species
is represented in the following woodcut, Fig.
XXV.; B, being a side view of the whole
flower, but with all the petals and sepals
excepting the labellum cut off, and A, being
a front view of the column. The upper sepal
and two upper petals surround and protect
the column; the two lower sepals project out

---

* I am much indebted to Mr. James Veitch of Chelsea for the
first specimen which I saw of this Orchid; subsequently Mr. S.
Rucker, so well known for his magnificent collection of Orchids,
generously sent me two fine spikes, and has aided me in the
kindest manner with other specimens.

at right angles. The flower stands more or
less inclined to either side, but with the
labellum downwards. The dull coppery and
orange-spotted tints,—the yawning chasm in
the great fringed labellum,—the one antenna
stuck out with the other hanging down—give
to these flowers a strange, lurid, and reptilian
appearance.

In front, in the middle of the column
(Fig. A), the deep stigmatic chamber (s)
may be seen; this is shown in the section
(Fig. XXVI. C), in which all the parts are
a little separated from each other, in order
that the mechanism may be made intelligible.
In the middle of the roof of the stigmatic
chamber, far back (d, in A), the upturned
anterior end of the viscid disc of the pollinium
may be discerned. The disc is continuous on
each side with a little fringe of membrane,
which joins the bases of the two antennæ.
Over the disc the protuberant heart-shaped
rostellum projects, and this is closely covered
by a thin membrane. This membrane is the
pedicel of the pollinium, with its lower end
attached (*ped* in sect. C, and in A) to the
superior surface of the viscid disc, and with

Fig. XXV.

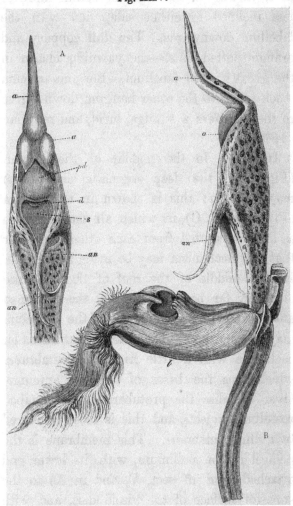

CATASETUM SACCATUM.

Fig. XXVI.

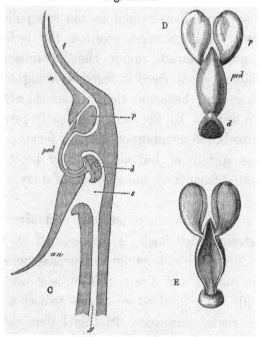

CATASETUM SACCATUM.

a. anther.	*l.* labellum.
an. antennæ of the rostellum.	*p.* pollen-masses.
d. disc of pollinium.	*pd* or *ped.* pedicel of pol-
f. filament of anther.	linium.
g. germen or ovarium.	*s.* stigmatic chamber.

A. Front view of column.
B. Side view of flower, with all the sepals and petals removed
  except the labellum.
C. Section through the column, with all the parts a little separated.
D. Pollinium, upper surface.
E. Pollinium, lower surface, which lies in contact with the
  rostellum.

L

its upper end running under the anther-
cells (*a*), and there united to the two pollen-
masses. In its natural position the pedicel
lies much bowed round the protuberant
rostellum; when freed it forcibly straightens
itself, and at the same time its lateral edges
curl inwards. In the bud, at an early period
of growth, the membranous pedicel forms part
of the rostellum, but subsequently becomes
separated from it by the solution of a layer of
cells.

The pollinium when let free, and after it
has straightened itself, is represented at D,
Fig. XXVI.; and its under surface, which
lies in contact with the rostellum, is shown at
E, with the lateral edges of the pedicel now
much curled inwards. In this latter view,
the clefts in the under sides of the two pollen-
masses are shown. Within the cleft, at its
base, a layer of strong extensible tissue is
attached, forming the caudicle, by which the
pollen-masses are united to the pedicel. The
lower end of the pedicel is joined to the disc
by a flexible hinge, which occurs in no other
genus, so that the pedicel can play backwards
and forwards as far as the upturned end

(Fig. D) of the disc permits. The disc is
large and thick; it consists of a strong upper
membrane, to which the pedicel is united,
with an inferior cushion of great thickness, of
pulpy, flocculent, and viscid matter. The
posterior margin (or lower margin in Fig. D)
is much the most viscid portion, and this
necessarily first strikes any object when the
pollinium is ejected. The viscid matter soon
sets hard. The whole surface of the disc
is kept damp before ejection, by resting
against the roof of the stigmatic chamber;
in the section (Fig. C) it is represented, like
the other parts, a little separated from the
roof.

The connective membrane of the anther
(a in all the figures) is produced into a spike,
which adheres loosely to the pointed end of
the column; this pointed end (f, Fig. C)
homologically is the filament of the anther.
The anther is thus shaped apparently to give
leverage, so that it may be easily torn off by
a pull at its lower end, when the pollinium
is let free and is jerked out by the elasticity
of its pedicel.

The labellum stands at right angles to the

column, or hangs a little downwards ; its
lateral and basal lobes are turned under the
middle portion, so that an insect can stand
only in front of the column. In the middle of
the labellum there is a deep cavity, bordered
by crests : this cavity does not secrete nectar,
but its walls are thick and fleshy, and have a
slightly sweet nutritious taste. I believe, as
we shall hereafter see, that insects visit the
flowers to gnaw these fleshy walls and crests.
The extremity of the left-hand antenna stands
immediately over the cavity, and would almost
certainly be touched by an insect visiting this
part of the labellum for any purpose.

The antennæ are the most singular organs
of the flower, and occur in no other genus.
They form rigid, curved horns, tapering to a
point. They consist of a narrow ribbon of
membrane, with the edges curled inwards so
as to touch, but the edges are not united;
hence each horn is tubular, like an adder's
fang, with a slit down one side. They are
composed of numerous, much elongated,
generally hexagonal cells, pointed at both
ends ; and these cells (like those in most of
the other tissues of the flower) have nuclei

with nucleoli. The antennæ are prolongations of the sides of the anterior face of the
rostellum. As the viscid disc is continuous
with the little fringe of membrane on each
side, and as this fringe is continuous with the
bases of the antennæ, these latter organs are
put in direct connection with the disc. The
pedicel of the pollinium passes between the
bases of the two antennæ. The antennæ are
not free for their whole length; but their
exterior edges, for a considerable space, are
firmly united to and blended with the margins
of the stigmatic chamber.

In all the flowers which I have examined,
gathered from three plants, both antennæ
occupied the same position; but though otherwise alike, they do not stand symmetrically.
The extreme part of the left-hand antenna
bends upwards (see Fig. B, in which the
position is shown plainer than in A), and at
the same time a little inwards, so that its tip
is medial and guards the entrance into the
pit of the labellum. The right-hand antenna
hangs downwards, with its tip turned a little
outwards; owing to this position, the crease
or furrow formed by its inflexed edges is

externally visible in this antenna; whilst in
the other it is hidden along the under side.
As we shall immediately see, the depending
right-hand antenna is almost paralysed, and is
apparently functionless.

Now for the action of the parts. When the
left-hand antenna of this species (or either
antenna of the two following species) is
touched, the edges of the upper membrane of
the disc, which are continuously united to the
surrounding surface, instantaneously rupture,
and the disc is set free. The highly elastic
pedicel then instantly flirts the heavy disc out
of the stigmatic chamber with such force, that
the whole pollinium is ejected, bringing away
with it the two balls of pollen, and tearing the
loosely attached spike-like anther from the
top of the column. The pollinium is always
ejected with its viscid disc foremost. I imitated
this action with a minute strip of whalebone,
slightly weighted at one end to represent the
disc; and by bending it round a cylindrical
object, gently holding at the same time the
upper end under the smooth head of a pin, to
represent the retarding action of the anther,
I then let the lower end suddenly free, and

the whalebone was pitched forward, like the
pollinium of the Catasetum, with the weighted
end foremost.

That the disc is first jerked out, I ascertained
by pressing with a scalpel on the middle of
the pedicel and by then touching the antenna;
instantaneously out came the disc, but, owing
to the pressure on the pedicel, the pollinium
was not ejected. Besides the spring from the
straightening of the pedicel, elasticity in a
transverse direction comes into play: if a quill
be split lengthways, and the half be forced
longitudinally on a too thick pencil, immediately the pressure is removed the quill
jumps off; and an analogous action takes
place with the pedicel of the pollinium, owing
to the sudden inward curling of its edges.
These combined forces suffice to eject the pollinium with considerable force to the distance
of two or three feet. Several persons have
told me that, when touching the flowers of
this genus in their hot-houses, the pollinia
have struck their faces. I touched the antennæ of C. callosum whilst holding the
flower at about a yard's distance from the
window, and the pollinium hit the pane of

glass, and adhered to the smooth vertical surface by its adhesive disc.

The following observations on the nature of the excitement which causes the disc to separate from the surrounding parts, include some made on the two succeeding species. Several flowers were sent me by post and railroad, and must have been much jarred, but had not exploded. I let two flowers fall from a height of two or three inches on the table, but the pollinia were not ejected. I cut off the ovarium close under the flower, and the sepals, and in some cases the thick labellum, with a crash by a pair of scissors; but this violence produced no effect. Nor did deep pricks in various parts of the column even within the stigmatic chamber. A blow, sufficiently hard to knock off the anther suddenly, causes the ejection of the pollinium, as occurred to me once by accident. Twice I pressed rather hard on the pedicel, and consequently on the underlying rostellum, without any effect. Whilst pressing on the pedicel, I gently removed the anther, and then the pollen-bearing end of the pollinium sprang up from its elasticity, and this movement caused the disc to separate. M. Ménière,* however, has stated that the anther case sometimes detaches itself, or can be gently detached, without the disc separating, and that then the pedicel swings downwards in front of the stigmatic chamber.

After trials made on fifteen flowers of three species, I find that no moderate degree of violence on any part of the flower, excepting the antennæ, produces any effect. But when the right-hand antenna of C. saccatum, or either antenna of the two following species, is touched, the pollinium is instantly ejected. The extreme tip and the whole length of the antennæ are sensitive. In one specimen of C. tridentatum a touch from a bristle sufficed; in five specimens of C. saccatum a gentle touch from a fine needle was necessary; but in four other specimens a slight blow was requisite. In C. tridentatum a stream of air and of cold water from a small pipe did not suffice; nor in any case did a touch from a human hair; so that the antennæ are less sensitive than the rostellum of Listera. Such extreme sensitiveness would indeed have been useless to the

* 'Bull. de la Soc. Bot. de France,' tom. i. 1854, p. 367.

plant, for we have reason to believe that the flowers are visited by bulky insects.

That the disc does not separate by the simple mechanical movement of the antennæ is almost certain; for the antennæ firmly adhere for a considerable space to the sides of the stigmatic chamber, and are thus immoveably fixed near their bases. The flowers in some cases, when they first arrived, were not sensitive, but after the spikes had stood for a day or two in water they became sensitive. Whether this was owing to fuller maturity or to the absorption of water, I know not. Two flowers of C. callosum, which were completely torpid, were immersed in tepid water for an hour; and then the antennæ became highly sensitive; this indicates that the cellular tissue of the antennæ must be turgid in order to receive and convey the effects of a touch; and it would lead to the suspicion that a vibration is conveyed along them; if this be so, the vibration must be of some special nature, for ordinary jars of manifold greater force do not cause the act of rupture. Two flowers placed in hot water, but not so hot as to scald my fingers, spontaneously ejected their pollinia. The loss of a plant, on which I intended to try other experiments, prevented my ascertaining whether drops or vapour of acrid fluids would act. From these latter facts it may be doubted whether it can be a vibration from the gentle touch of a needle which is conveyed along the antennæ. In C. tridentatum I found that the antennæ were one inch and one-tenth of an inch in length, and a gentle touch from a bristle on the extreme tip was conveyed, as far as I could perceive, instantaneously throughout this length. I measured the length of several cells in the tissue composing the antennæ, and on a rough average it appeared that the stimulus must travel through no less than from seventy to eighty closed cells.

We may, at least, safely conclude that the antennæ, which are characteristic of the genus Catasetum, are specially adapted to receive and convey the effects of a touch to the disc of the pollinium; causing the membrane to rupture, and the whole pollinium to be ejected by its elasticity. If we required further proof, nature has afforded it in the case of the so-called genus Monachanthus, which, as we shall see, is the female plant of Catasetum tridentatum, and

has no pollinia to eject, and here the antennæ
are entirely absent.

I have stated that in C. saccatum the right-
hand antenna invariably hangs down, with
the tip turned slightly outwards, and that it is
almost paralysed. I ground my belief on five
trials, in which I violently hit, bent, and
pricked this antenna, and produced no effect;
but immediately afterwards touching the left-
hand antenna with much less force, the pollinium was shot forth. In a sixth case a forcible
blow on the right-hand antenna did cause the
act of ejection, so that it is not completely
paralysed. As this antenna does not guard
the labellum, which seems in all Orchids to be
the part attractive to insects, its sensitiveness
would be useless.

From the large size of the flower, more
especially of the viscid disc, and from its
wonderful power of adhesion, we may safely
infer that the flowers are visited by large
insects. The viscid matter sticks so firmly
when it has set hard, and the pedicel is so
strong (though very thin and only one-
twentieth of an inch in breadth at the hinge),
that to my surprise it supported for a few

seconds a weight of 1262 grains, that is nearly three ounces; and it supported for a considerable time a slightly less weight. When the pollinium is shot forth, the large spike-like anther is generally carried with it. When the disc strikes a flat surface like a table, the momentum from the weight of the anther often carries the pollen-bearing end beyond the disc, and the pollinium is thus affixed in a wrong direction, supposing it to have been attached to an insect's body, for the fertilisation of another flower. The flight is also often rather crooked.* But it must not

* M. Baillon ('Bull. de la Soc. Bot. de France,' tom. i. 1854, p. 285) states that Catasetum luridum ejects its pollinia always in a straight line, and in such a direction that it sticks fast to the bottom of the concavity of the labellum: when in this position he imagines that it fertilises the flower in a manner not clearly explained. In a subsequent paper in the same work (p. 367) M. Ménière justly disputes M. Baillon's conclusion. He remarks that the anther-case is easily detached, and sometimes naturally detaches itself; in this case the pollinia swing downwards by the elasticity of the pedicel, and the viscid disc still remains attached to the roof of the stigmatic chamber. M. Ménière then hints that, by the subsequent and progressive retraction of the pedicel, the pollen-masses might be carried into the stigmatic chamber. This is not possible in the three species which I have examined, and would be useless. But M. Ménière himself then goes on to show how important insects are to the fertilisation of Orchids; and apparently he infers that their agency comes into play with Catasetum, and that this plant does not fertilise itself. Both M.

be forgotten that under nature the ejection
is caused by the antennæ being touched by a
large insect standing on the labellum, which
will thus have its head and thorax placed
near to the anther. A rounded object thus
held is always accurately struck in the
middle, and, when removed with the pollinium
adhering to it, the weight of the anther depresses the hinge of the pollinium; and in
this position the anther readily drops off,
leaving the balls of pollen free and in a
proper position for the act of fertilisation.
The utility of so forcible an ejection may be
to drive the soft and viscid cushion of the
disc against the hairy thorax of a large hymenopterous insect, or the sculptured thorax
of a flower-feeding beetle. When attached,
assuredly no force which the insect could exert
would remove the disc and pedicel; but the
caudicles are ruptured without much difficulty,
and thus the balls of pollen would be left on
the viscid stigmatic surface of a female flower.

Baillon and M. Ménière correctly describe the curved position in
which the elastic pedicel lies before it is set free. Neither of
these botanists seems to be aware that the species of Catasetum
(at least the three which I have examined) are exclusively male
plants.

Catasetum callosum.—This species* is smaller than the last, but resembles it in most respects. The edge of the labellum is covered with papillæ ; the pit in its middle is small, and behind it there is an elongated anvil-like projection, — facts which I mention from the relation of this form to Myanthus barbatus, presently to be described. The yellow-coloured pedicel is much bowed, and is joined by a hinge to the extremely viscid disc. When either antenna is touched, the pollinia are ejected with much force. The two antennæ stand symmetrically on each side of the anvil-like projection, with their tips lying within the pit of the labellum. The walls of this pit have a pleasant nutritious taste. The antennæ are remarkable, from their whole surface being roughened with papillæ. The plant is a male.

Catasetum tridentatum.—The general appearance of this species, which is very different from that of the two former species, is represented by Fig. XXVII., with a sepal on each side cut off.

---

* A fine spike of flowers of this species was most kindly sent me by Mr. Rucker, and was named for me by Dr. Lindley.

Fig. XXVII.

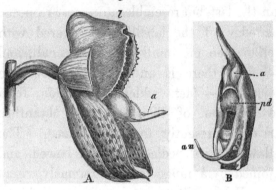

CATASETUM TRIDENTATUM.

a. anther.	*an.* antennæ.
pd. pedicel of pollinium.	*l.* labellum.

A. Side view of flower in its natural position with the (properly) lower sepals cut off.
B. Front view of column, placed upright.

The flower stands with the labellum uppermost, that is in a reversed position compared with most Orchids. The labellum forms a helmet or bucket, with the distal portion represented by three minute points. It is clear from its position that the labellum cannot hold nectar; but its walls are thick, and have, as in the other species, a pleasant nutritious taste. The stigmatic chamber, though functionless as a stigma, is of large

size. The summit of the column, and the
spike-like anther, are not so much elongated
as in C. saccatum. In other respects there
is no important difference. The antennæ are
of greater length; their tips for about one-
twentieth of their length are roughened by
cells produced into papillæ.

The pedicel of the pollinium is articulated
as before by a hinge to the disc; the anterior
end of the disc is upturned, so that, when
attached to an insect's head, the pedicel cannot move backwards, only forwards,—a movement which apparently comes into play in
the fertilisation of the female plant. The
disc is, as in the other species, of large size,
and its posterior end, which during ejection
first strikes any object, is much more viscid
than the rest of the surface. This latter
surface is drenched with a milky fluid, which
rapidly turns brown when exposed to the
air, and sets into a cheesy consistence. The
upper surface of the disc consists of strong
membrane formed of polygonal cells, each
containing one or several balls of brown
translucent matter. This membrane rests
on, and adheres to, a thick cushion, formed

of rounded balls of brown matter (which lower down in the cushion become extremely irregular in shape) separated from each other, and embedded in transparent, structureless, highly elastic matter. This cushion towards the posterior end of the disc passes into the extremely viscid matter. This latter matter when consolidated is brown, translucent, and homogeneous. Altogether the disc presents a much more complex structure than in the other Vandeæ.

I need not further describe this species, excepting the position of the antennæ. They occupied exactly the same position in the six flowers examined. They do not stand symmetrically. They are both sensitive,—whether in an equal degree I will not say. They both lie curled within the bucket-like labellum; the left-hand one stands higher up, with its inwardly bowed extremity in the middle; the right-hand antenna lies lower down and crosses the whole base of the labellum, with its tip just projecting beyond the left margin of the base of the column. From the position of the petals and sepals, an insect visiting the flower would almost certainly alight on the

crest of the labellum; but it could hardly
gnaw any part of the great cavity of the
labellum without touching one of the two
antennæ, for the left-hand one guards the
upper part, and the right-hand one the
lower part; and when these are touched
the pollinium will infallibly be ejected and
strike the head or thorax of the insect.

The position of the antennæ in this Catasetum may be compared with that of a man
with his left arm raised and bent so that his
hand stands in front of his chest, and with
his right arm crossed lower down so that the
fingers project just beyond his left side. In
Catasetum callosum both arms are held lower
down, and are extended symmetrically. In
C. saccatum the left arm is bowed and held
in front, as in the C. tridentatum, but rather
lower down; whilst the right arm hangs
down almost paralysed, with the hand turned
a little outwards. In every case notice will
be given in an admirable manner, when
an insect visits the labellum, and the time
has at last arrived for the ejection of the
pollinium and for its transportal to the female
plant.

Catasetum tridentatum is interesting under another point of view. Botanists were astonished when Sir R. Schomburgk * stated that he had seen three forms, believed to constitute three distinct genera, namely, Catasetum tridentatum, Monachanthus viridis, and Myanthus barbatus, all growing on the same plant. Lindley remarked † that "such cases shake to the foundation all our ideas of the stability of genera and species." Sir R. Schomburgk affirms that he has seen hundreds of plants of C. tridentatum in Essequibo without ever finding one specimen with seeds,‡ but that he was surprised at the gigantic seed-

* 'Transactions of the Linnæan Soc.' vol. xvii. p. 522. Another account by Dr. Lindley has appeared in the 'Botanical Register,' fol. 1951, of a distinct species of Myanthus and Monachanthus appearing in the same scape: he alludes also to other cases. Some of the flowers were in an intermediate condition, which is not surprising, seeing that in diœcious plants we sometimes have a partial resumption of the characters of both sexes. Mr. Rodgers of Riverhill informs me that he imported from Demerara a Myanthus, but that when it flowered a second time it was metamorphosed into a Catasetum. Dr. Carpenter ('Comparative Physiology,' 4th edit. p. 633) alludes to an analogous case which occurred at Bristol.

† The 'Vegetable Kingdom,' 1853, p. 178.

‡ Brongniart states ('Bull. de la Soc. Bot. de France, tom. ii. 1855, p. 20) that M. Neumann, a skilful fertiliser of Orchids, could never succeed in fertilising Catasetum.

vessels of the Monachanthus ; and he correctly remarks that " here we have traces of sexual difference in Orchideous flowers."

From what I had myself previously observed, I was led to examine carefully the female organs of C. tridentatum, callosum, and saccatum. In no case was the stigmatic surface viscid, as it is in all other Orchids (except as we shall see in Cypripedium), and as is indispensable for the securing the pollen-masses by the rupture of the caudicles : I carefully looked to this point both in young and old flowers of C. tridentatum. When the surface of the stigmatic chamber and of the stigmatic canal of the above-named three species is scraped off, after having been kept in spirits, it is found to be composed of utriculi, with nuclei of the proper shape, but not nearly so numerous as with ordinary Orchids. The utriculi cohere more together and are more transparent; I examined for comparison the utriculi of many kinds of Orchids which had been kept in spirits, and in all found them much less transparent. In C. tridentatum, the ovarium is shorter, much less deeply furrowed, narrower at the base, and

internally more solid than in the Monachanthus. Again, in all three species of Catasetum the ovule-bearing cords are short; and
the ovules present a considerably different
appearance, in being thinner, more transparent, and less pulpy than in the numerous
other Orchids examined for comparison.
They were, however, in not so completely
an atrophied condition as in Acropera. Although they correspond so closely in general
appearance and position with true ovules,
perhaps I have no strict right so to designate
them, as I was unable in any case to make
out the opening of the testa and the included
nucleus; nor were the ovules ever inverted.

From these several facts, namely,—the
shortness, smoothness, and narrowness of the
ovarium, the shortness of the ovule-bearing
cords, the state of the ovules themselves, the
stigmatic surface not being viscid, the empty
condition of the utriculi,—and from Sir R.
Schomburgk never having seen C. tridentatum producing seed in its native home, we
may confidently look at this species, as well as
the other two species of Catasetum, as male
plants.

Fig. XXVIII.

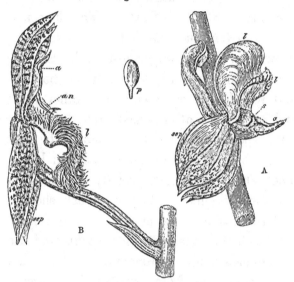

MYANTHUS BARBATUS.	MONACHANTHUS VIRIDIS.
a. anther.	*p.* pollen-mass, rudimentary.
an. antennæ.	*s.* stigmatic cleft.
l. labellum.	*sep,* two lower sepals.

A. Side view of Monachanthus viridis in its natural position.
   (The shading in both drawings has been added from
   M. Reiss' drawing in the Linnæan Transactions.')
B. Side view of Myanthus barbatus in its natural position.

With respect to Monachanthus viridis and
Myanthus barbatus, the President and officers
of the Linnæan Society have kindly permitted
me to examine the spike bearing these two
flowers, preserved in spirits, and sent home

by Sir R. Schomburgk. They are here represented (Fig. XXVIII.). The flower of
the Monachanthus, like that of the Catasetum,
grows lower side uppermost. The labellum
is not nearly so deep, especially on the sides,
and its edge is crenated. The other petals
and sepals are all reflexed, and are not so
much spotted as in the Catasetum. The bract
at the base of the ovarium is much larger.
The whole column, especially the filament
and-the spike-like anther, are much shorter;
and the rostellum is much less protuberant.
The antennæ are entirely absent, and the
pollen-masses are rudimentary. These are
interesting facts, from corroborating the view
taken of the function of the antennæ; for as
there are no proper pollinia to eject, there
could be no use in an organ to convey the
stimulus from a touch to the rostellum. I
could find no trace of a viscid disc or pedicel;
if they exist, they must be extremely rudimentary, for there is hardly any space for the
embedment of the disc.

Instead of a large stigmatic chamber, there
is a narrow transverse cleft close beneath the
small anther. I was able to insert one of the

pollen-masses of the male Catasetum into this cleft, which from having been kept in spirits was lined with coagulated beads of viscid matter, and with utriculi. The utriculi, differently from those in Catasetum, were charged (after having been kept in spirits) with brown matter. The ovarium is longer, thicker near the base, and more plainly furrowed than in Catasetum; the ovule-bearing cords are also much longer, and the ovules more opaque and pulpy, as in all common Orchids. I believe that I saw the opening at the partially inverted end of the testa, with a large nucleus projecting; but as the specimens had been kept many years in spirits and were somewhat altered, I dare not speak positively. From these facts alone it is almost certain that Monachanthus is a female plant; and Sir R. Schomburgk saw it seeding abundantly. Altogether this flower differs in a most remarkable manner from that of the male Catasetum tridentatum, and it is no wonder that they were formerly ranked as distinct genera.

The pollen-masses offer so curious and good an illustration of a structure in a rudimentary condition that they are worth description;

but first I must briefly describe the perfect
pollen-masses of the male Catasetum. These
may be seen at D and E, Fig. XXVI. (p. 217),
attached to the pedicel : they consist of a large
sheet of cemented or waxy pollen-grains, folded
over so as to form a sack, with an open slit (E)
along the lower surface ; into this slit cellular
tissue enters whilst the pollen is in the course
of development. Within the lower and produced end of each pollen-mass a layer of highly
elastic tissue, forming the caudicle, is attached;
the other end being attached to the pedicel of
the rostellum. The exterior grains of pollen
are more angular, have thicker walls, and are
yellower than the interior grains. In the
early bud the two pollen-masses are enveloped
in two conjoined membranous sacks, which are
soon penetrated by the two produced ends of
the pollen-masses and by their caudicles; and
then the ends of the caudicles adhere to the
pedicel. Before the flower expands the membranous sacks including the two pollen-masses
open, and leave them resting naked on the
back of the rostellum.

In Monachanthus the two membranous sacks
containing the rudimentary pollen-masses on

the contrary never open; they easily separate from each other and from the anther. The tissue of which they are formed is thick and pulpy. Like most rudimentary parts, they vary greatly in size and in form. The included and therefore useless pollen-masses are not one-tenth of the bulk of the pollen-masses of the male; they are flask-shaped (*p*, p. 239, Fig. XXVIII.), with the lower and produced end greatly exaggerated and almost penetrating through the exterior or membranous sack. The flask is closed, and there is no fissure along the lower surface. The exterior pollen-grains are square and have thicker walls than the interior grains, just as in the proper male pollen; and, what is very curious, each cell has its nucleus. Now, R. Brown has stated * that in the early stages of the formation of the pollen-grains in ordinary Orchids a minute areola or nucleus is often visible; so that the rudimentary pollen-grains of the Monachanthus apparently have retained—as is so general with rudiments in the animal kingdom—an embryonic character. Lastly, at the base, within the flask of pollen, there

* 'Transactions of the Linnæan Soc.' vol. xvi. p. 711.

is a little mass of brown elastic tissue,—that is, a vestige of a caudicle,—which runs far up the pointed end of the flask, but does not (at least in some of the specimens) come to the surface, and could not have been attached to any part of the rostellum. These rudimentary caudicles are, therefore, utterly useless.

We thus see that every single detail of structure which characterises the male pollen-masses is represented, with some parts exaggerated and some parts slightly modified, by the mere rudiments in the female plant. Such cases are familiar to every observer, but can never be examined without renewed interest. At a period not far distant, naturalists will hear with surprise, perhaps with derision, that grave and learned men formerly maintained that such useless organs were not remnants retained by the principle of inheritance at corresponding periods of early growth, but were specially created and arranged in their proper places like dishes on a table (this is the comparison of a distinguished naturalist) by an Omnipotent hand " to complete the scheme of nature."

We now come to the third form, Myanthus

barbatus (Fig. XXVIII., B), often borne on
the same plant with the two preceding
forms. Its flower, in external appearance,
but not in essential structure, is the most
different of all three. It generally stands in
a reversed position, compared with Catasetum
and Monachanthus, that is, with the labellum
downwards. The labellum is fringed in an
extraordinary manner with long papillæ; it
has a quite insignificant medial cavity, at
the hinder margin of which a curious curved
and flattened horn projects. The other petals
and sepals are spotted and elongated, with
the two lower sepals alone reflexed. The
antennæ are not so long as in the male
C. tridentatum, and they project symmetrically on each side of the horn-like projection
at the base of the labellum, with their tips,
which are not roughened with papillæ, almost
entering the medial cavity. The stigmatic
chamber is of nearly intermediate size between that of the male and female forms:
it is lined with utriculi charged with brown
matter. The straight and well-furrowed
ovarium is nearly twice as long as in Monachanthus, but is not so thick where it joins

the flower; the ovules are not so numerous as in the female form, but are opaque and pulpy after having been kept in spirits, and resemble them in all respects. I believe, but dare not, as in the case of the Monachanthus, speak positively, that I saw the nucleus projecting from the testa. The pollinia are about a quarter of the size of those of the male Catasetum, but have a perfectly well developed disc and pedicel. The pollen-masses were lost in the specimens examined by me; but fortunately Mr. Reiss has given, in the Linnæan Transactions, a drawing of them, showing that they are of due proportional size and have the proper folded or cleft structure : so that there can hardly be a doubt that they were functionally perfect. As we thus see that both male and female organs are apparently perfect, Myanthus barbatus may be considered as the hermaphrodite form of the same species, of which the Catasetum is the male and Monachanthus the female.

It is not a little singular that the hermaphrodite Myanthus should resemble in its whole structure much more closely the male form of two distinct species (namely, C.

saccatum, and more especially C. callosum) than either its own male or female form.

Finally, the genus Catasetum is interesting in an unusual degree in several respects. The separation of the sexes is unknown in other Orchids, excepting as we shall see probably in the allied genus Cycnoches and in the before-given case of Acropera. In Catasetum we have three sexual forms, generally borne on separate plants, but sometimes mingled together ; and these three forms are wonderfully different from each other, much more different than, for instance, a peacock is from a peahen. But the appearance of these three forms on the same plant now ceases to be an anomaly, and can no longer be viewed as an unparalleled instance of variability.

Still more interesting is this genus in its mechanism for fertilisation. We see a flower patiently waiting with its antennæ stretched forth in a well-adapted position, ready to give notice whenever an insect puts its head into the cavity of the labellum. The female Monachanthus, not having pollinia to eject, is destitute of antennæ. In the male and hermaphrodite forms, namely Catasetum tri-

dentatum and Myanthus, the pollinia lie
doubled up, like a spring, ready to be instantaneously shot forth when the antennæ
are touched; the disc end is always projected
foremost, and is coated with viscid matter
which quickly sets hard and firmly affixes
the hinged pedicel to the insect's body. The
insect flies from flower to flower, till at last
it visits a female or hermaphrodite plant: it
then inserts one of the masses of pollen into
the stigmatic cavity. When the insect flies
away the elastic caudicle, made weak enough
to yield to the viscidity of the stigmatic surface, breaks, and leaves behind the pollen-
mass; then the pollen-tubes slowly protrude,
penetrate the stigmatic canal, and the act of
fertilisation is completed. Who would have
been bold enough to have surmised that the
propagation of a species should have depended
on so complex, so apparently artificial, and
yet so admirable an arrangement?

I have seen two other genera belonging
to the sub-family of Catasetidæ, namely, Mormodes and Cycnoches; but the latter arrived
in a broken condition.

Mormodes ignea.—To show how difficult it sometimes is to understand the manner of fertilisation of Orchids, I may state that I had carefully examined twelve flowers,* trying various experiments and recording the results, before I could at all make out the meaning and action of the several parts. It was plain that the pollinia were ejected, as in Catasetum, but how each part of the flower played its proper part I could not even conjecture. I had given up the case as hopeless, until, summing up my observations, the explanation presently to be given, and subsequently proved by repeated experiments to be correct, suddenly occurred to me.

The flower presents an extraordinary appearance, and its mechanism is more curious even than its appearance (Fig. XXIX.). The base of the column is bent backwards, at right angles to the ovarium or footstalk, and then resumes an upright position to near its summit, where it is again bent. It is, also, twisted in a unique manner, so that its front

* I must express my cordial thanks to Mr. Rucker, of West Hill, Wandsworth, for having lent me a plant of this Mormodes with two fine spikes, bearing an abundance of flowers, and for having allowed me to keep the plant for a considerable time.

M 3

Fig. XXIX.

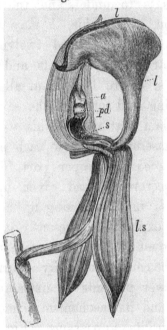

MORMODES IGNEA.

Lateral view of flower, with the upper sepal and the near upper petal cut off.

N.B. The labellum in the drawing is a little lifted up, to show the depression in its surface, which ought to be pressed close down on the bent summit of the column.

a. anther.	*l.* labellum.
pd. pedicel of pollinium.	*l s.* lower sepal.
s. stigma.	

surface, including the anther, rostellum, and the upper part of the stigma, faces laterally to either the right or left hand in the flowers on the opposite sides of the spike. The twisted stigmatic surface, in an equally peculiar manner, extends down to the base of the column : at its upper end it forms a deep cavity, beneath the protuberant rostellum (*pd* in the drawing), in which the large viscid disc of the pollinium is lodged.

The anther-case (*a* in the drawing) is elongated and triangular, closely resembling that of Catasetum : it does not extend up to the apex of the column. This apex consists of a thin flattened filament, which in the bud is straight, but before the flower expands becomes much bent by the pressure of the labellum. A group of spiral vessels runs up the column as far as the summit of the anther-case ; they are then reflexed and run some way down the anther-case. The point of reflexion forms a short thread-like hinge by which the notched top of the anther-case is articulated to the column close beneath its bent summit. The hinge, although less than

a pin's head in size, is of paramount importance, for it is sensitive and conveys the
stimulus from a touch to the disc of the
pollinium : it serves also as a guide in the
act of ejection. As it conveys the necessary
stimulus to the disc, one may suspect that
part of the tissue of the rostellum, which
lies in close contact with the filament of the
anther, runs up to this point ; but I could
detect no difference in structure in comparison with the same parts in Catasetum.
The cellular tissue round the hinge is gorged
with fluid, for a large drop exudes when the
anther is torn off during the ejection of the
pollinium. This gorged condition may perhaps facilitate the ultimate rupture of the
hinge.

The pollinium does not differ much from
that of Catasetum (see Fig. XXVI. D, p. 217) :
it lies curved round the rostellum, which
is less protuberant than in that genus.
The upper and broad end of the pedicel,
however, extends beneath the pollen-masses,
and they are attached by rather weak caudicles to a medial crest on its upper surface.

The viscid surface of the large disc lies

in contact with the roof of the stigmatic cavity, so that it cannot be touched by insects. The anterior end of the disc is furnished with a small dependent curtain (dimly shown in Fig. XXIX.); and this, before the act of ejection, is continuously joined on each side to the upper margins of the stigmatic depression. The pedicel is united to the posterior end of the disc; but when the disc is freed, the lower part of the pedicel becomes doubly bent, so that it then appears as if attached to the centre of the disc by a hinge.

The labellum is truly remarkable : it is narrowed at its base into a nearly cylindrical footstalk, and its sides are so much reflexed as almost to meet at the back. After rising up perpendicularly it forms an arch, over and behind the summit of the column, against which it is firmly pressed. The labellum at this point (even in the bud) is depressed into a slight cavity, which receives the ulti- mately bent summit of the column. This slight depression manifestly represents the large hollow, with thick fleshy walls, in the labellum of the several species of Catasetum

and other Vandeæ, which serves to attract
insects; here, by a singular change of function, it merely keeps the labellum in its
proper position on the summit of the column.
In the drawing (Fig. XXIX.) the labellum is
represented as forcibly raised so as to show
the depression and the bent filament. The
labellum in its natural position may almost
be compared to a huge cocked-hat, supported
by a footstalk and placed on the head of the
column.

The twisting of the column, which I have
seen in no other Orchid, causes all the important organs of fructification to face to the
left in the flowers on the left side of the
spike, and to face to the right in all those
on the right side. So that two flowers taken
from opposite sides of the spike and held in
the same relative position are seen to be
twisted in opposite directions. One single
flower, which was crowded by the others, was
barely twisted, so that its column faced the
labellum. The labellum is also slightly
twisted: for instance, in the flower figured
which faces to the left, the midrib of the
labellum first bends to the right hand, and

then back again, but in a lesser degree, to the left, so as to press against the posterior surface of the summit of the column. The twisting of all the parts of the flower commences in the bud.

The position thus acquired by the several organs is of the highest importance; for if the column and labellum had not been twisted laterally, the pollinia, when shot forth, would have struck the overarching labellum and been thrown back, as actually occurred with the single abnormal flower having a nearly straight column. If all the organs had not been twisted in opposite directions on the two sides of the crowded spike, so as to face always to the outside, there would not have been a clear space for the ejection of the pollinia and their adhesion to insects.

When the flower is mature the three sepals hang down, but the two upper petals remain nearly upright. The bases of the sepals, and especially of the two upper petals, are thick and swollen and have a yellowish tint; they are so gorged, when quite mature, with fluid, that, when punctured by a fine glass tube, the fluid rises to some height in it by capillary attraction. These swollen bases, as well as the footstalk of the labellum, have a decidedly sweet and pleasant taste; and I can hardly doubt that they are attractive to insects, for no free nectar is secreted.

I will now endeavour to show how all the parts of the flower are co-ordinated and act together. The pollinium, as in Catasetum, lies bowed round the rostellum; in that genus when freed it merely straightens itself with force, in this Mormodes something more takes place. If the reader will look forward to Fig. XXX., p. 267, he will see a section of the flower-bud of another species of Mormodes, which differs only in the shape of the anther and in the viscid disc having a much deeper dependent curtain. Now let him suppose the pedicel of the pollinium to be so elastic that when freed it not only straightens itself, but suddenly bends backwards with a reversed curvature, so as to form an irregular hoop; he will then see that the exterior surface of the curtain, which is not viscid, will lie on the anther-case, and the viscid surface of the disc will be on the outside of the hoop.

This is exactly what takes place with our
present Mormodes. But the pollinium assumes with such force its reversed curvature
(aided, apparently, by a transverse curling
outwards of the two margins of the pedicel),
that it instantaneously rebounds from the
protuberant face of the rostellum. As the
two pollen - masses adhere, at first, rather
firmly to the anther-case, the latter is torn
off by its base by the rebound; and as the
thin hinge at the summit of the anther-case
does not at first break, the pollinium with
the anther-case is instantly swung upwards
like a pendulum : but in the course of the
upward swing the hinge yields and the
whole body is projected perpendicularly up in
the air, an inch or two above and close in
front of the terminal part of the labellum.
When no object is in the way, and the
pollinium falls down, it generally alights and
sticks, though not firmly, in the fold on the
crest of the labellum, directly over the column.
I witnessed repeatedly all that has been here
described.

The curtain of the disc, which, after the
pollinium has formed itself into a hoop, lies

on the anther-case, is of considerable service in preventing the viscid matter from the disc adhering to the anther, and thus permanently retaining the pollinium in the form of a hoop. This would have been fatal, as we shall presently see, to a subsequent movement in the pollinium necessary to the fertilisation of the flower. In some of my experiments, when the free action of the parts was checked, this did occur, and the pollinium, with the anther, remained glued together in the shape of an irregular hoop.

I have already stated that the minute hinge by which the anther-case is articulated to the column, a little way beneath its bent filamentary apex, is sensitive to a touch. I tried four times and found that I could touch with some force any other part; but when I gently touched this point with the finest needle, instantaneously the membrane which unites the disc to the edges of the cavity in which it is lodged, ruptured, and the pollinium was shot upwards and alighted on the crest of the labellum as just described.

Now let us suppose an insect to alight on the crest of the labellum (and no other con-

venient landing-place is afforded), and then
to lean over in front of the column to gnaw
or suck the bases of the petals swollen with
sweet fluid. The weight and movements of
the insect would press and move the labellum
together with the bent underlying summit of
the column; and the latter, pressing on the
hinge in the angle, would cause the ejection
of the pollinium, which would infallibly
strike the head of the insect and adhere to
it. I tried by placing my gloved finger
on the summit of the labellum, with the tip
just projecting beyond its margin, and then
gently moving my finger it was really beautiful to see how instantly the pollinium
was projected upwards, and how accurately
the whole viscid surface of the disc struck
my finger and firmly adhered to it. Nevertheless, I doubt whether the weight and
movement of an insect would suffice to thus
act indirectly on the sensitive point; but look
at the drawing and see how probable it is
that an insect leaning over would place its
front legs over the edge of the labellum on
the summit of the anther-case, and would
thus touch the sensitive point; the pollinium

would then be ejected, and the viscid disc
would certainly strike and adhere to the
insect's head.

Before proceeding, it will be worth while
to mention some of the early trials which I
made. I pricked deeply the column in different parts and the stigma, and cut off the
petals, and even the labellum, without causing
the ejection of the pollinium; once, however,
in cutting rather roughly through the thick
footstalk of the labellum, the ejection ensued,
no doubt caused by the filamentary summit of
the column being disturbed. When I gently
prised up the anther-case at its base or at
one side, the pollinium was ejected, but then
the sensitive hinge must necessarily have
been bent. When the flower has long
remained expanded and is nearly ready for
spontaneous ejection, a slight jar on any part
of the flower causes the action. Pressure on
the thin pedicel, and therefore on the underlying protuberant rostellum at the basal edge
of the anther-case, is followed by the ejection
of the pollinium; but this is not surprising,
as the stimulus from a touch on the sensitive
hinge has to be conveyed through this part

of the rostellum to the disc. In Catasetum slight pressure on this point does not cause the ejection; but in this genus the protuberant part of the rostellum does not lie in the line of conveyed stimulus between the antennæ and the disc. A drop of chloroform, of spirits of wine, and of boiling water placed on this part of the rostellum produced no effect; nor, to my surprise, did exposure of the whole flower to vapour of chloroform.

Seeing that this part of the rostellum was sensitive to pressure, and that the flower was laterally widely open, and being preoccupied with the case of Catasetum, I at first felt convinced that insects entered the lower part of the flower and touched the rostellum. Accordingly I pressed the rostellum with variously-shaped objects, but the viscid disc never once well adhered to the object. If I used a thick needle, the pollinium, when ejected, formed a hoop round it with the viscid surface outside; if I used a broad flat object, the pollinium struggled against it and sometimes coiled itself up spirally, but the disc either did not adhere at all or very imperfectly. At the close of

the twelfth trial I was in despair. The
strange position of the labellum, perched on
the summit of the column, ought to have
shown me that here was the place for experiment. I ought to have scorned the notion
that the labellum was thus placed for no
good purpose : I neglected this plain guide,
and for a long time completely failed to
understand the structure of the flower.

We have seen that when the pollinium is
freely ejected upwards it adheres by the
whole viscid surface of the disc to any object
projecting beyond the edge of the labellum
directly over the column. When thus attached, it forms an irregular hoop, with the
torn-off anther-case still covering the pollen-
masses and lying in close contact with the
disc, but protected from adhering to it by the
curtain. Whilst in this position the projecting and bowed pedicel would effectually prevent the pollen-masses from being placed on
the stigma of a flower, even supposing the
anther-case to have fallen off. Now let us
suppose the pollinium to be attached to an
insect's head, and observe what takes place.
The pedicel when first separated by the act

of ejection from the rostellum is quite wet
on its inferior surface : as this surface dries,
the pedicel slowly straightens itself, and when
perfectly straight the anther-case readily drops
off. The pollen-masses are now naked, and
they are attached by easily ruptured caudicles
to the end of the pedicel, at the right distance
and on that surface which would naturally
be placed in contact with the viscid stigma
when the insect visited another flower, so
that every detail of structure is now perfectly
well adapted for the act of fertilisation.

When the anther-case drops off, it has
performed its triple function ; namely, its
hinge as an organ of sense, its weak attachment to the column as a guide causing the
pollinium to be at first swung perpendicularly
upwards, and its lower margin, together with
the curtain of the disc, as a protection to the
pollen-masses from being permanently glued
to the viscid disc.

From observations made on fifteen flowers,
the straightening of the pedicel takes from
about twelve to fifteen minutes. The first
movement causing the act of ejection is due
to elasticity ; the second slow movement,

which counteracts the first, is due to the drying of the outer and convex surface; but this
movement differs from that observed in the
pollinia of so many Vandeæ and Ophreæ,
for, when the pollinium of this Mormodes
is placed in water, it does not recover the
hoop-like form which it had acquired by
elasticity.

Mormodes ignea is an hermaphrodite. The
pollinia are well developed. The curiously
elongated stigmatic surface is extremely viscid
and abounds with innumerable utriculi, the
contents of which shrink and become coagulated after immersion for less than an hour
in spirits of wine. When placed in spirits
for a day, the utriculi are so acted on that
they disappear, and this I have noticed in
no other Orchid. The ovules, after exposure
to spirits for a day or two, present the usual
semi-opaque, pulpy appearance common to
all hermaphrodite and female Orchids. From
the unusual length of the stigmatic surface
I expected that, if the pollinia were not
ejected from the excitement of a touch, the
anther-case would detach itself, and that the
pollen-masses would swing downwards and

fertilise their own flower. Accordingly, I left four flowers untouched; after remaining expanded from eight to ten days, the elasticity of the pedicel conquered the force of attachment and the pollinia were spontaneously ejected, but they were always wasted.

Although Mormodes ignea is an hermaphrodite, yet it must be, in fact, as truly diœcious as Catasetum as far as the concurrence of two individuals is concerned in the act of reproduction ; for, as it takes from twelve to fifteen minutes, after the act of ejection, before the pedicel of the pollinium straightens itself and the anther-case drops off, it is almost certain that an insect with the pollinium attached to its head would within this time leave one plant and fly to another.

A second species of Mormodes, sent me, unnamed, by Mr. Veitch, is very different from M. ignea both in general appearance and in structure. The yellow petals and sepals are reflexed; the thick labellum is singularly shaped, with its upper surface convex, like a shallow basin turned upside

down. The thin column is of extraordinary
length, and arches over the labellum. The
appearance of the flower is like that of
Cycnoches represented on the cover of this
volume; but the column in the drawing is too
much foreshortened.

The specimens unfortunately arrived broken;
but some large flower-buds, of which a section
is here given, show the essential structure.
We see the elastic pedicel of the pollinium
bowed as in the last species; but at the period
of growth here represented, the pedicel was
still united to the rostellum, the future line of
separation being shown by a layer of hyaline
tissue, indistinct towards the upper end of the
disc. The disc is of gigantic size, and its lower
end is produced into a great fringed curtain,
which hangs in front of the stigmatic chamber.
The adhesion of this disc, when mature, to any
object is surprisingly strong. The margins
of the stigmatic chamber on each side are
slightly protuberant; and these protuberances, like the antennæ in Catasetum, are
continuous with the rostellum. The anther
is widely different in shape from that in the
last species and in Catasetum, and apparently

would retain the pollen-masses with more force.

Fig. XXX.

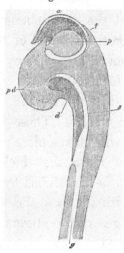

SECTION OF THE FLOWER-BUD OF A MORMODES.

a. anther.	*d.* disc of pollinium.
f. filament of anther.	*s.* stigmatic chamber.
p. pollen-mass.	*g.* stigmatic canal leading to
pd. pedicel of pollinium, barely separated from the rostellum.	the ovarium.

From information given me by Mr. Veitch it appears that, when the end of the column is touched, some movement takes place; and, from the analogy of Catasetum, it is probable that the protuberances on the margin of the

N 2

stigmatic chamber are sensitive. These protuberances may be provisionally considered as
nascent antennæ. It is obvious, from the section given, that the disc, as long as its viscid
surface rests against the roof of the stigmatic
chamber, cannot adhere to any object; but
Mr. Veitch informs me that, when he has
touched the end of the column, the disc has
adhered to his fingers. These facts perhaps
suffice to show what takes place under nature.
A large insect visits the thick and fleshy
labellum, which is overarched by the column,
and touches with its back the sensitive and
protuberant edges of the stigmatic chamber:
when thus excited the disc alone is flirted out
and adheres to the insect's back; the insect
flies away, drags the pollen-masses from
under the anther, and carries them to another
flower. The insect, standing in the same
position, inserts the pollen-masses into the
stigmatic chamber, and the flower is fertilised.

When I received the flowers of this second
species of Mormodes and of Cycnoches, I did
not examine their ovaria and stigmas, for I
knew nothing at that time about the sexual

forms of Catasetum. Mormodes ignea is an hermaphrodite; if the second species is an hermaphrodite, it cannot be fertilised, owing to the size of the curtain of the disc, until its own pollinium has been removed. With respect to Cycnoches, it is known, from an account published by Lindley,* that C ventricosum produces on the same scape flowers with a simple labellum, others with a much segmented labellum, and others in an intermediate condition. From the analogous differences in the labellum of the sexes in Catasetum, we may believe that we here see the male, female, and hermaphrodite forms of the Cycnoches.

Of Lindley's fourth and sixth Tribes, viz. the Ophreæ and Neotteæ, plenty of British forms have been described. Of the fifth Tribe, the Arethuseæ, I have not seen any living flowers. Judging from statements with respect to three distantly allied forms in this tribe, mechanical aid is requisite for their fertilisation. Irmisch makes this remark

* 'Vegetable Kingdom,' 1853, p. 177. Lindley has also published in the 'Botanical Register,' fol. 1951, a similar case of the production of two forms on the scape in another species of Cycnoches.

with respect to the Epipogium aphyllum.*
Mr. Rodgers, of Sevenoaks, informs me that
in his hot-house species of Limodorum did not
set their fruit without aid ; and this is likewise well known to be the case with the
Vanilla. This latter genus is cultivated for
its aromatic pods in Tahiti, Bourbon, and
the East Indies ; but does not fruit without
artificial aid.† This fact shows that some
insect in its own American home is specially
adapted for its fertilisation ; and that the
insects of the above-named tropical regions,
where the Vanilla flourishes, either do not
visit the flowers, though they secrete an
abundance of nectar, or do not visit them in
the proper method.

We have now arrived at Lindley's last
and seventh Tribe, including only one genus,
Cypripedium, which differs from all other
Orchids far more than any other two do
from each other. An enormous amount of

* 'Beiträge zur Biologie der Orchideen,' 1853, s. 55.
† For Bourbon see 'Bull. Soc. Bot. de France,' tom. i. 1854,
p. 290. For Tahiti see H. A. Tilley, 'Japan, the Amour, &c.,'
1861, p. 375. For the East Indies see Morren in Annals and
Mag. of Nat. Hist., 1839, vol. iii. p. 6.

extinction must have swept away a multitude of intermediate forms, and left this single genus, now widely disseminated, as a record of a former and more simple state of the great Orchidean Order. Cypripedium possesses no rostellum; all three stigmas being fully developed, but confluent. That anther, which is present in all other Orchids, is here rudimentary, and is represented by a singular shield-like projecting body, deeply notched or hollowed out on its lower margin. There are two fertile anthers which belong to an inner whorl, represented in ordinary Orchids by various rudiments. The pollen-grains do not consist of three or four united granules, as in all other genera, excepting the degraded Cephalanthera. The grains are not united into waxy masses, nor tied together by elastic threads, nor furnished with a caudicle. The labellum is of large size, and as in all other Orchids is a compounded organ.

The following remarks apply only to the four species which I have seen, namely, C. barbatum, purpuratum, insigne, and venustum. The manner of fertilisation is here widely different from that in all the many

before-given cases. The labellum is folded
round the short column, so that its edges
nearly meet along the dorsal surface; and
its broad extremity is folded over and backwards in a peculiar manner, so as to form a

Fig. XXXI.

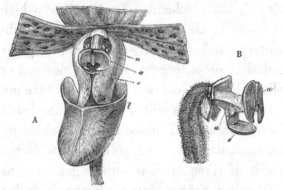

CYPRIPEDIUM.

a. anther.	*s.* stigma.
a'. rudimentary, shield-like anther.	*l.* labellum.

A. Flower viewed from above, exhibiting the dorsal surface, with
    the sepals and petals, excepting the labellum, partly cut
    off. The labellum is slightly depressed, so that the dorsal
    surface of the stigma is brought outside; the edges of the
    labellum have thus become a little separated, with the toe
    depressed.
B. Side view of column, with all the sepals and petals removed.

sort of shoe, which closes up the end of the
flower. Hence arises the English name of
Ladies'-slipper. In the position in which the

flower grows, and as it is here represented,
the dorsal surface, with the edges of the
labellum almost meeting, is uppermost. The
stigmatic surface is slightly protuberant, and
is not viscid; it fronts the basal surface of
the labellum. The margin of the upper and
dorsal surface of the stigma can be barely distinguished between the edges of the labellum
and through the notch in the rudimentary,
shield-like anther(a') ; but in the drawing
(s, Fig. A) this margin is brought outside the
edges of the labellum by their depression ; the
toe also of the labellum is a little bent down,
so that the flower is represented as rather
more open than it really is. The edges of the
pollen-masses of the two lateral anthers (a)
can be seen lying low down within the labellum and projecting a little beyond the column.
The grains of pollen are immersed in, and
coated by, viscid fluid, which is so glutinous
that it can be drawn out into threads. As
the two anthers stand behind and above
the lower convex surface (see Fig. B) of the
stigma, it is impossible that the glutinous
pollen can get on to this, the fertile surface,
without mechanical aid.

An insect could reach the extremity of the labellum, or the toe of the slipper, through the longitudinal dorsal slit; but according to all analogy the basal portion in front of the stigma would be the most attractive part. Now, as the flower is closed at the end, owing to the toe of the labellum being upturned, and as the dorsal surface of the stigma, together with the large shield-like rudimentary anther, almost close the basal part of the medial slit, two convenient passages alone are left for an insect to reach with its proboscis the lower part of the labellum; namely, directly over and close outside the two lateral anthers. If an insect were thus to act, and it could hardly act in any other way, it would infallibly get its proboscis smeared with the glutinous pollen, as I found to occur with a bristle thus inserted. When the bristle smeared with pollen was pushed further into the flower, especially if pushed in by the little notch outside the anther, some of the glutinous pollen was generally left on the slightly convex stigmatic surface. The proboscis of an insect would effect this latter operation better than a bristle, owing to its flexibility and power of movement. Thus an insect would place either the flower's own pollen on to the stigma, or, flying away, would carry the pollen to another flower. Which of these two contingencies commonly occurs, will depend on whether the insect first inserts its proboscis directly over the anther, or outside by the little notch.

We thus see how important, or rather how necessary for the fertilisation of the plant, is the curious slipper-like shape of the labellum, in leading insects to insert their probosces by the lateral passages close to the anthers. The upper, shield-like, rudimentary anther is equally, and in the same manner, necessary.

The economy shown by nature in her resources is striking : in all Orchids seen by me, excepting Cypripedium, the stigma is more or less concave and is viscid, so as to secure the dry pollen, brought to it by means of the viscid matter secreted by the modified stigma, called the rostellum. In Cypripedium alone the pollen is glutinous, and takes on itself the function of viscidity, which in other Orchids belongs to both the true stigma and

the modified stigma or rostellum. The stigma
itself, on the other hand, in Cypripedium entirely loses its viscidity, and at the same time
becomes slightly convex, so as more effectually
to rub off the glutinous pollen adhering to an
insect's proboscis. Thus the act of fertilisation is completed, and there is no superfluity
in the means employed.

SECRETION OF NECTAR.

Many exotic Orchids secrete plenty of
nectar in our hot-houses. I have found the
horn-like nectaries of Aerides filled with
fluid ; and Mr. Rodgers, of Sevenoaks, informs me that he has taken crystals of sugar
of considerable size from the nectary of A.
cornutum. In nearly all the flowers of
Angræcum distichum sent me from Kew,
insects had bitten holes through the nectaries,
so as to get more readily at the nectar : if
insects were invariably to follow this bad
habit in the plant's native African home,
undoubtedly it would soon become extinct,
for it would never produce a seed. The
nectar-secreting organs present great diver-

sity in structure and position in the various genera; but apparently always form part of the base of the labellum. In Dendrobium chrysanthum the nectary consists of a shallow saucer; in Evelyna of two large united cellular balls; in Bolbophyllum cupreum of a medial furrow. In Cattleya the nectary penetrates the ovarium; in Angræcum sesquipedale it attains, as we have seen, the astonishing length of above eleven inches; but I will not detail the several cases. The nectar-secreting apparatus of Coryanthes, however, is so remarkable, as described by M. Ménière,* that it must not be quite passed over: two little horns near the strap-like junction of the labellum with the base of the column, secrete so much limpid nectar, with a slightly sweet taste, that it slowly drops down. M. Ménière estimates the total quantity secreted by a single flower at about an English ounce. But the remarkable point is that the deeply-hollowed end of the labellum hangs some way down, exactly beneath the two little horns, and catches the drops as

* 'Bulletin de la Soc. Bot. de France,' tom. ii. 1855, p. 351.

they fall, just like a bucket suspended some way beneath a dripping spring of water.

Although the secretion of nectar is of the highest importance to Orchids by attracting insects, which are indispensable to their fertilisation, yet it would seem that the secretion acts also, at least in some cases, as an excretion. One may suspect this to be the case with the Coryanthes from the immense quantity secreted; but the bracteæ of some Orchids have been observed * to secrete nectar; and as these lie outside the flower, the secretion cannot serve any useful purpose by attracting insects. Mr. Rodgers informs me that he has observed much nectar secreted at the base of the footstalks of the flowers in Vanilla. It is in perfect accordance with the scheme of nature, as worked out by natural selection, that matter excreted to free the system from superfluous or injurious substances

---

* J. G. Kurr, 'Ueber die Bedeutung der Nektarien,' 1833, s. 28, on the authority of Treviranus and Curt. Sprengel. The calyx of certain species of Iris also (idem, s. 25) secretes nectar. I have observed much nectar secreted by the stipules of Vicia sativa and faba, which is eagerly collected by bees. Glands on the under side of the leaves of the common laurel also secrete nectar, which, though the drops are extremely minute, is sought for by various insects.

should be utilised for purposes of the highest importance. To give an example in strong contrast with flowers and honey, the larvæ of certain beetles (Cassidæ, &c.) use their own excrement to make an umbrella-like protection for their tender bodies.

In Cypripedium the slipper-like labellum seems well adapted to hold and collect nectar; but I have found no such collection in the four previously named species; nor does C. calceolus ever secrete nectar according to Kurr.* The labellum, however, in these four species is studded with hairs; and I have almost always noticed on their tips little drops of slightly viscid fluid, which if sweet would certainly suffice to attract insects; this viscid fluid when dried forms a brittle crust, but I could not perceive in it any traces of crystallisation.

It may be remembered that in the first chapter evidence was given proving that nectar is never found within the spur-like nectary of several species of Orchis, but that between its two membranes there is an abundant supply of fluid. In all the species

* 'Bedeutung der Nektarien,' 1833, s. 29.

thus characterised the viscid matter of the
disc of the pollinium sets hard in a minute
or two, and it would be an advantage to the
plant if an insect were delayed in getting the
nectar by having to puncture the nectary at
several points, so that time should be allowed
for the setting of the viscid matter. On the
other hand, in all the Ophreæ which have
nectar ready stored within the nectary the
viscid matter does not rapidly set hard, and
there would be no advantage in insects being
delayed.

In the case of cultivated exotic Orchids
which have a nectary, without nectar, it is of
course impossible to feel sure that it would be
empty under more natural conditions. Nor
have I made many comparative observations
on the rate of setting of the viscid matter of
the disc in exotic forms. Nevertheless it seems
that certain Vandeæ are in the same predica-
ment as our British species of Orchis; thus
Calanthe masuca has a very long nectary,
which in all the specimens examined was
internally quite dry, and was inhabited by
powdery Cocci; but in the intercellular spaces
between its two coats there was much fluid;

in this species the viscid matter of the disc, after its surface had been disturbed, entirely lost its adhesiveness in two minutes. In an Oncidium the disc, similarly disturbed, became dry in one minute and a half; in an Odontoglossum in two minutes; neither of these Orchids have free nectar. On the other hand, in Angræcum sesquipedale, which has free nectar stored within the lower end of its nectary, the disc of the pollinium, removed from the plant, and with its surface disturbed, was strongly adhesive after forty-eight hours.

Sarcanthus teritifolius offers a more curious case. The disc quite lost its viscidity, after the removal of the pollinium from the rostellum, in less than three minutes. Hence it might have been expected that fluid would have occurred in the intercellular spaces of the nectary, but none within the nectary; nevertheless there was fluid in both places, so that we here see the two conditions of the nectar-secreting organs combined in the same flower. It might perhaps have been thought that insects would have rapidly sucked the free nectar and neglected that between the two coats. But I strongly suspect that insects are

delayed in sucking the free nectar, so as to
allow the viscid matter to set hard, by totally
different means. In this Orchid the labellum
with its nectary is an extraordinary organ. I
wished to have had a drawing of its structure ;
but found that it was as hopeless as to give
one of the wards of a complicated lock : even
the skilful Bauer, with numerous figures and
sections on a large scale, hardly makes the
structure intelligible. So complicated is the
passage that I failed in repeated attempts to
pass a bristle from the outside of the flower
into the nectar-receptacle ; or in a reversed
direction from the cut-off end of the nectar-
receptacle to the outside. No doubt an insect
with a voluntarily flexible proboscis could
wind it through the passages, and thus get
the nectar ; but in effecting this, some delay
would be caused ; and time would be allowed
for the curious square viscid disc to become
securely cemented to the insect's head or body.

As in Epipactis the cup at the base of
the labellum serves as a nectar-receptacle, I
expected to find that the analogous cup in
Stanhopea, Acropera, &c., would serve for the
same purpose ; but I could never find a drop

of nectar in it. According, also, to M. Ménière,* this is never the case in these genera, nor in Gongora, Cirrhæa, and others. In Catasetum tridentatum, and in its female the Monachanthus, we see that the upturned cup cannot possibly serve as a nectar-receptacle. What then attracts insects to these flowers? That they must be attracted is certain ; more especially in the case of Catasetum, in which the sexes stand on separate plants. In many genera of Vandeæ there is no trace of any nectar-secreting organ or receptacle; but in all these cases, as far as I have seen, the labellum is either thick and fleshy or is furnished with excrescences. The labellum, for instance, of Oncidium and of Odontoglossum offers all sorts of singular protuberances. In Calanthe we have (Fig. XXIV.) a cluster of odd little spherical warts on the labellum, together with an extremely long nectary, which does not include nectar ; in Eulophia viridis the short nectary is in the same condition, and the labellum is covered with longitudinal and fimbriated ridges. In some also of the Ophreæ, which have no nectary, the labellum, as in the Fly Orphys, has two

* 'Bulletin Bot. Soc. de France,' tom. ii. 1855, p. 352.

shining protuberances at its base, placed beneath the two pouches. Lindley has remarked that the use of these strange and diversified excrescences is quite unknown.

From the position, relatively to the viscid disc of the pollinium, which these excrescences hold, and from the absence of nectar, it seems to me highly probable that they afford food, and thus attract either Hymenoptera or flower-feeding Coleoptera. I mention this belief because a close examination of the flowers of the Vandeæ, which in their native country have had their pollinia removed, would soon settle this point. There is no more inherent improbablity in a flower being habitually fertilised by an insect coming to feed on the labellum, than in seeds being habitually disseminated by birds attracted by the sweet pulp in which they are embedded. But I am bound to state that Dr. Percy had the thick and furrowed labellum of a Warrea analysed for me, by fermentation over mercury, and it gave no evidence of containing more saccharine matter than the other petals. On the other hand, the thick labellum of Catasetum, and even the bases of the upper petals in Mormodes ignea, had, as previously stated, a slightly sweet, rather pleasant, and nutritious taste.

We have now done with exotic Orchids. To me the study has been most interesting of these wonderful and often beautiful productions, so unlike common flowers, with all their many adaptations, with parts capable of movement, and other parts endowed with something so like, though no doubt really different from, sensibility. The flowers of Orchids, in their strange and endless diversity of shape, may be compared with the great vertebrate class of Fish, or still more appropriately with tropical Homopterous insects, which seem to us in our ignorance as if modelled by the wildest caprice.

CHAPTER VII.

Homologies of Orchid-flowers — The great modification which they have undergone — Gradation of organs, of the rostellum, of the pollen-masses — Formation of the caudicle — Genealogical affinities — Mechanism of the movement of the pollinia — Uses of the petals — Production of seed — Importance of trifling details of structure — Cause of the vast diversity of structure for the same general purpose — Cause of the perfection of the contrivances in Orchids — Summary on insect-agency — Nature abhors perpetual self-fertilisation.

THE theoretical structure of few flowers has been so largely discussed as that of Orchids; nor is this surprising, seeing how unlike they are to common flowers. No group of organic beings can be well understood until their homologies are made out; that is, until the general pattern, or, as it is often called, the ideal type, of the several members of the group is intelligible. No one member may now exist exhibiting the full pattern; but this does not make the subject less important to the naturalist,—probably makes it more important for the full understanding of the group.

The homologies of any being, or group of beings, can be most surely made out by tracing their embryological development when that is possible; or by the discovery of organs in a rudimentary condition; or by tracing, through a long series of beings, a close gradation from one part to another, until the two parts, or organs, employed for widely different functions, and most unlike each other, can be joined by a succession of short links. No instance is known of a close gradation between two organs, unless they be homologically one and the same organ.

The importance of the science of Homology rests on its giving us the key-note of the possible amount of difference in plan within any group; it allows us to class under proper heads the most diversified organs; it shows us gradations which would otherwise have been overlooked, and thus aids us in our classification; it explains many monstrosities; it leads to the detection of obscure and hidden parts, or mere vestiges of parts, and shows us the meaning of rudiments. Besides these practical uses, to the naturalist who believes in the gradual modification of organic beings, the science of Homology clears away the mist from such terms as the scheme of nature, ideal types, archetypal patterns or ideas, &c. ; for these terms come to express real facts. The naturalist, thus guided, sees that all homologous parts or organs, however much diversified, are modifications of one and the same ancestral organ; in tracing existing gradations he gains a clue in tracing, as far as that is possible, the probable course of modification during a long line of generations. He may feel assured that, whether he follows embryological development, or searches for the merest rudiments, or traces gradations between the most different beings, he is pursuing the same object by different routes, and is tending towards the knowledge of the actual progenitor of the group, as it once grew and lived. Thus the subject of Homology gains largely in interest.

Although this subject, under whatever aspect it be viewed, will always be most interesting to the student of nature, it is very doubtful whether the following details on the homological nature of the flowers of Orchids will be endured by the general reader. If, indeed, he should care to see how much light, though far from perfect, homology throws on a subject, this will, perhaps, be nearly as good an instance as could be given. He will see how curiously a flower may be moulded out of many separate organs,—how perfect the cohesion of primordially distinct parts may become,—how organs may be used for purposes widely different from their proper function,—how other organs may be entirely suppressed, or leave mere useless emblems of their former existence. Finally, he will see how enormous has been the total amount of change from the simple parental or typical structure which these flowers have undergone.

Robert Brown first clearly discussed the homologies of Orchids,* and left, as might be expected, little to be done. Guided by the general structure of monocotyledonous plants, and by various considerations, he propounded the doctrine that the flower properly consists of three sepals, three petals, six anthers in two whorls or circles (of which only one anther belonging to the outer whorl is perfect in all

* I believe his latest views are given in his celebrated paper, read Nov. 1-15, 1831, in the ' Linnæan Transactions,' vol. xvi. p. 685.

common forms), and of three pistils, with one
of them modified into the rostellum. These
fifteen organs are arranged as usual, alternately, three within three, in five whorls.
Of the existence of three of the anthers in
two whorls, R. Brown offers no sufficient
evidence, but believes that they are combined
with the labellum, whenever that organ presents crests or ridges. In these views Brown
is followed by the greatest living authority
on Orchids, namely, Lindley.

Brown traced the spiral vessels in the
flower by making transverse sections,* and
only occasionally, as far as it appears, by
longitudinal sections. As spiral vessels are
developed at a very early period of growth,
which always gives much value to an organ
in making out homologies; and as they are
apparently of high functional importance,
though their function is not well known, it

* ' Linn. Transact.' vol. xvi. p. 696-701. Link in his ' Bemerkungen über der Bau der Orchideen ' (' Botanische Zeitung,' 1849,
s. 745) seems to have also trusted to transverse sections. Had
he traced the vessels upwards I cannot believe that he would
have disputed Brown's view of the nature of the two anthers in
Cypripedium. Brongniart in his admirable paper (' Annales des
Sciences Nat.' tom. xxiv. 1831) incidentally shows the course of
some of the spiral vessels.

appeared to me, guided also by the advice of Dr. Hooker, to be worth while to trace upwards all the spiral vessels from the six groups surrounding the ovarium. Of the six ovarian groups of vessels, I will call (though not correctly) that under the labellum the anterior group; that under the upper sepal the posterior group; and the two groups on both sides of the ovarium the antero-lateral and postero-lateral groups.

The result of my dissections is given in the following diagram. The fifteen little circles represent so many groups of spiral vessels, in every case traced down to one of the six large ovarian groups. They alternate in five whorls, as represented; but I have not attempted to give the actual distances at which they stand apart. In order to guide the eye, the three central groups running to the three pistils are connected by a triangle.

Five groups of vessels run into the three sepals and two upper petals; three enter the labellum; and seven run up the great central column. These vessels are arranged, as may be seen, in rays proceeding from the axis of the flower; and all on the same ray invari-

ably run into the same ovarian group : thus
the vessels supplying the upper sepal, the

Fig. XXXII.

Upper or posterior sepal.

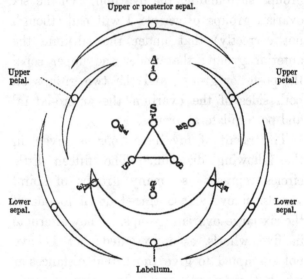

Labellum.

Section of the flower of an orchid.
The little circles show the position of the spiral vessels·

S S. Stigmas ; S$_r$, stigma modified into the rostellum.
A$_1$. Fertile anther of the outer whorl ; A$_2$ A$_3$, anthers of the same
    whorl combined with the lower petal, forming the labellum.
a$_1$ a$_2$. Rudimentary anthers of the inner whorl (fertile in Cypripe-
    dium), generally forming the clinandrum; a$_3$, third anther of
    the same whorl, when present, forming the front of the column.

fertile anther (A 1), and the upper pistil or
stigma (*i. e.* rostellum S r), all unite and form
the posterior ovarian group. Again, the

vessels supplying one of the lower sepals, the
corner of the labellum, and one of the two
stigmas (S), unite and form the antero-lateral
group ; and so with all the other vessels.

Hence, if the existence of groups of spiral
vessels can be trusted, and Dr. Hooker informs
me that he has never known them to speak
falsely, the flower of an Orchid certainly
consists of fifteen organs, in a much modified
and confluent condition. We see three stigmas,
with the two lower ones generally confluent,
and with the upper one modified into the
rostellum. We see six stamens, arranged in
two whorls, with one alone (A 1) generally
fertile. In Cypripedium, however, two
stamens of the inner whorl (*a* 1 and *a* 2) are
fertile, and in other Orchids these two are
represented in various ways more plainly
than the remaining stamens. The third stamen
of the inner whorl (*a* 3), when its vessels
can be traced, forms the front of the column :
Brown thought that it often formed a medial
excrescence, or ridge, cohering to the labellum ;
or, in the case of Glossodia,* a filamentous

---

* See Brown's observations under Apostasia in Wallich's
' Plantæ Asiaticæ rariores,' 1830, p. 74.

organ, freely projecting in front of the labellum. The former conclusion does not agree with my dissections ; about Glossodia I know nothing. The two infertile stamens of the outer whorl (A 2, A 3) were believed by Brown to be sometimes represented by lateral excrescences on the labellum; I find these vessels invariably present in the labellum of every Orchid examined,—even when the labellum is very narrow, or quite simple, as in Malaxis, Herminium, and Habenaria.

We thus see that an Orchid-flower consists of five simple parts, namely, three sepals and two petals; and of two compounded parts, namely, the column and labellum. The column is formed of three pistils, and generally of four stamens, all completely confluent. The labellum is formed of one petal and two petaloid stamens of the outer whorl, likewise completely confluent. I may remark, as making this fact more probable, that in the allied Marantaceæ the stamens, even the fertile stamens, are often petaloid, and partially cohere. This view of the nature of the labellum explains its large size, its frequently tripartite form, and especially its manner of coherence to the column, unlike that of the other petals.* As rudimentary organs vary much, we can thus also probably understand the variability, which Dr. Hooker informs me is characteristic of the excrescences on the labellum. In some Orchids which have a spur-like nectary, the two sides are apparently formed by the two modified stamens; thus in Gymnadenia conopsea (but not in Orchis pyramidalis), the vessels, proceeding from the antero-lateral ovarian group, run down the sides of the nectary; those from the anterior group run down the exact middle of the nectary, then returning up the opposite side form the mid-rib of the labellum. The extension of these lateral elements of the nectary apparently explains the tendency, as in Calanthe, Orchis morio, &c., to the bifurcation of its extremity.

The number, position, and course of the spiral vessels exhibited in the diagram (Fig. XXXII.), were observed in some Vandeæ and Epidendreæ.† In the Malaxeæ all were

* Link remarks on the manner of coherence of the labellum to the column in his 'Bemerkungen' in 'Bot. Zeitung,' 1849, s. 745.

† It may be advisable to give a few details on the flowers which I dissected; but I looked to special points, such as the course

observed excepting *a* 3, which is the most
difficult of all to trace, and which apparently

of the vessels in the labellum, in many cases not worth here
giving. In the Vandeæ I traced all the vessels in *Catasetum
tridentatum* and *saccatum*; the great group of vessels going to the
rostellum separate (as likewise in Mormodes) from the posterior
ovarian group, beneath the bifurcation supplying the upper sepal
and fertile anther; the anterior ovarian group runs a little way
along the labellum before bifurcating, and sending a group (a_3) up
the front of the column; the vessels proceeding from the postero-
lateral group run up the back of the column, on each side of
those running to the fertile anther, and do not go to the edges of
the clinandrum. In *Acropera luteola* the base of the column.
where the labellum is attached, is much produced, and the vessels
of the whole anterior ovarian group are similarly produced; those
(a_3) going up the front of the column are abruptly reflected
back; the vessels at the point of reflexion are curiously hardened,
flattened, and produced into odd crests and points. In an Oncidium
I traced the vessels S_r to the viscid gland of the pollinium. In the
Epidendreæ I traced all the vessels in a Cattleya; and all in *Evelyna
carivata* except a_3, which I did not search for. In the Malaxeæ
I traced all in *Liparis pendula* except a_3, which I do not believe
is present. In *Malaxis paludosa* I traced nearly all the vessels.
In *Cypripedium barbatum* and *purpuratum* I traced all except a_3,
which I am nearly sure does not exist. In the Neotteæ I traced
in *Cephalanthera grandiflora* all the vessels, excepting that to the
aborted rostellum and those to the two auricles a_1 and a_2, which
were certainly absent. In *Epipactis* I traced all excepting a_1, a_2,
and a_3, which are certainly absent. In *Spiranthes autumnalis*
the vessel S_r runs to the bottom of the fork of the rostellum:
there are no vessels to the membranes of the clinandrum in this
Orchid nor in *Goodyera*. In none of the Ophreæ do the vessels a_1,
a_2, and a_3 occur. In *Orchis pyramidalis* I traced all the others,
including two to the two separate stigmas: in this species the
contrast between the vessels of the labellum and of the other
sepals and petals is striking, as in the latter the vessels do not

is oftenest absent. In the Cypripedeæ, again, all were traced except a 3, * which, I feel pretty sure, was here really absent : in this tribe the stamen (A 1) is represented by a conspicuous shield-like rudiment, and a 1 and a 2 are developed into two fertile anthers. In the Ophreæ and Neotteæ all were traced, with branch, whilst the labellum has three vessels, the lateral ones running of course into the antero-lateral ovarian group. In *Gymnadenia conopsea* I traced all the vessels; but I am not sure whether the vessels supplying *the sides* of the upper sepal do not, as in the allied Habenaria, wander from their proper course and enter the postero-lateral ovarian group: the vessel S_r, going to the rostellum, enters the little folded crest of membrane, which projects between the bases of the anther-cells. Lastly, in *Habenaria chlorantha* I traced all the vessels, excepting of course the three of the inner staminal whorl, and I looked carefully for a_3 : the vessel supplying the fertile anther runs up the connective membrane between the two anther-cells, but does not bifurcate : the vessel to the rostellum runs up to the top of the shoulder or ledge beneath the connective membrane of the anther, but does not bifurcate and extend to the two widely-separated viscid discs.

* From Irmisch's (' Beiträge zur Biologie der Orchideen,' 1853, s. 78 and 42) description of the development of the flower-bud of Cypripedium, it would appear that there is a tendency to the formation of a free filament in front of the labellum, as in the case of Glossodia before mentioned ; and this will perhaps account for the absence of spiral vessels, proceeding from the anterior ovarian group, and coalescing with the column. In Uropedium, a genus which A. Brongniart (' Annal. des. Sc. Nat.,' 3rd series, Bot. tom. xiii. p. 114) considers closely allied to, and even perhaps a monstrosity of, Cypripedium, a third fertile anther occupies this same exact position.

the important exception of the vessels belonging to the three stamens (a 1, a 2, and a 3)
of the inner whorl. In Cephalanthera grandiflora, however, I clearly saw a 3 proceeding
form the anterior ovarian group, and running
up the front of the column; this anomalous
member of the Neotteæ has no rostellum, and
the vessel marked S r in the diagram was
entirely absent, though seen in every other
Orchid.

Although in no true Orchid, excepting Cypripedium, the two anthers (a 1 and a 2) of
the inner whorl are fully developed, their
rudiments are generally present and are often
utilised; for they generally form the membranous sides of the clinandrum, or cup on
the summit of the column, which includes
and protects the pollen - masses. These
rudiments thus aid their fertile brother-
anther. In the young flower-bud of Malaxis
paludosa, the close resemblance between
the membranes of the clinandrum and the
fertile anther, in shape, texture, and in the
height to which the spiral vessels extend, is
most striking: it is impossible to doubt that
in these two membranes we see two rudimen-

tary anthers. In Evelyna, one of the Epidendreæ, the clinandrum was similarly formed,
as were the horns of the clinandrum in
Masdevallia, which likewise serve to keep the
labellum at the proper distance from the
column. In Liparis pendula and some other
forms, these two rudimentary anthers formed
not only the clinandrum, but likewise wings,
projecting on each side of the entrance to the
stigmatic cavity, and serving as guides for the
insertion of the pollen-masses. In Acropera
and Stanhopea, as far as I could make out,
the membranous borders of the column, down
to its base, were also thus formed ; but in
other cases, as in Cattleya, the wing-like
borders of the column seemed to be simple
developments of the two pistils. In this latter
genus, as well as in Catasetum, these same two
rudimentary stamens, judging from the position of the vessels, served chiefly to strengthen
the back of the column ; and the strengthening
of the front of the column is the sole function
of the third stamen $a\,3$) of the inner whorl, in
those cases in which I observed it. This third
stamen runs up the middle of the column to
the lower edge, or lip, of the stigmatic cavity.

I have said that in the Ophreæ and Neot-
teæ the spiral vessels, marked *a* 1, *a* 2, *a* 3 in
the diagram, are entirely absent, and I looked
carefully for them; but in nearly all the
members of these two tribes, two small papillæ,
or auricles as they have been often called,
stand in exactly the position which the two
first of these three anthers would have occupied, had they been developed. Not only
do they stand in this position, but the column
in some cases, as in Cephalanthera, has on
each side a prominent ridge, running from
them to the bases or mid-ribs of the two upper
petals; that is, in the proper position of the
filaments of these two stamens. It is, again,
impossible to doubt that the membranes of
the clinandrum in Malaxis are formed by
these two anthers in a rudimentary and
modified condition. Now, from the perfect
clinandrum of Malaxis, through that of Spiranthes, Goodyera, Epipactis latifolia, and
E. palustris (see Fig. XIV., p. 103, and XIII.,
p. 94), to the minute and slightly flattened
auricles in the genus Orchis, a perfect gradation can be traced. Hence I conclude that
these auricles are doubly rudimentary; that

is, that they are rudiments of the membranous sides of the clinandrum, these membranes themselves being rudiments of the two anthers so often referred to. The absence of spiral vessels running to the auricles by no means seems sufficient to overthrow these several arguments on their much-disputed nature ; that such vessels may quite disappear, we have proof in Cephalanthera grandiflora, in which the rostellum and its vessels are completely aborted.

Finally, then, with respect to the six stamens or anthers which ought to be represented in every Orchid : the three belonging to the outer whorl are always present, with the upper one generally fertile, and the two lower ones invariably petaloid and forming part of the labellum ; the three stamens of the inner whorl are less plainly developed, especially the lower one, a 3, which, when it can be detected, serves only to strengthen the column, and, in some rare cases, according to Brown, forms a separate projection, or filament ; the upper two anthers of this inner whorl are fertile in Cypripedium, and in other cases are generally represented either by mem-

branous expansions, or by minute auricles
without spiral vessels. These auricles, however, are sometimes quite absent, as in some
species of Ophrys.

On this view of the homologies of Orchid-
flowers, we can understand the existence of
the conspicuous central column, — the large
size, generally tripartite form, and peculiar
manner of attachment of the labellum,—the
origin of the clinandrum,—the relative position of the single fertile anther in most
Orchids, and of the two fertile anthers in
Cypripedium,—the position of the rostellum,
as well as of all the other organs, — and,
lastly, the frequent occurrence of a bilobed
stigma, and the occasional occurrence of two
distinct stigmas.

I have encountered only one case of difficulty on the foregoing views, namely in Habenaria, and in the allied Bonatea. These
flowers have undergone such an extraordinary
amount of distortion, owing to the wide separation of their anther-cells and of the two
viscid discs of the rostellum, that any anomaly
in them is the less surprising. The anomaly
relates only to the vessels supplying the sides

of the upper sepal and of the two upper petals; the vessels running into their mid-ribs and into all the other more important organs pursue the same identical course as in all the other Ophreæ. The vessels on the sides of the upper sepal, instead of uniting with the mid-rib, and entering the posterior ovarian group, diverge and enter the postero-lateral groups : again, the vessels on the anterior side of the upper petals, instead of uniting with the mid-rib and entering the postero-lateral ovarian groups, diverge, or wander from their proper course, and enter the antero-lateral groups.

This anomaly is so far of importance, as it throws some doubt on the view which I have taken of the labellum being always an organ compounded of one petal and two petaloid stamens; for if any one were to assume that from some unknown cause the lateral vessels of the lower petal in an early progenitor of the Orchidean order had wandered from their proper course into the antero-lateral ovarian groups, and that this structure had been inherited by all existing Orchids, even by those with the smallest and simplest labellums, I could answer only as follows; but the answer is, I think, satisfactory. From the analogy of other monocotyledonous plants, we might expect the hidden presence of fifteen organs in the flowers of Orchids, arranged alternately in five whorls; and in Orchid-flowers we do find fifteen groups of vessels exactly thus arranged. Hence there is a strong probability that the vessels, A 2 and A 3, which enter the sides of the labellum, not in one or two cases, but in all the Orchids seen by me, and which occupy the precise position which they would have occupied had they supplied two normal stamens, do really represent modified and petaloid stamens, and are not lateral vessels of the lower petal which have wandered from their proper course. In Habenaria and Bonatea,* on the other hand,

---

* In Bonatea speciosa, of which I have examined only dry specimens sent me by Dr. Hooker, the vessels from the sides of the upper sepal enter the postero-lateral ovarian group, exactly as in Habenaria. The two upper petals are divided down to their bases, and the vessels supplying the anterior segment and those supplying the *anterior portion* of the posterior segment unite and then run, as in Habenaria, into the antero-lateral (and therefore wrong) groups. The anterior segments of the two upper petals cohere with the labellum, making it, in a most unusual manner, five-segmented. The two wonderfully protuberant stigmas also cohere to the upper surface of the labellum; and the lower sepals apparently also cohere to its under side.

the vessels from the sides of the upper sepal and of the two upper petals, which enter the wrong ovarian groups, cannot possibly represent any now lost and once distinct organs.

We have now finished with the general homologies of the flowers of Orchids. It is interesting to look at one of the magnificent exotic species, or, indeed, at one of our humblest forms, and observe how profoundly it has been modified, as compared with all ordinary flowers,—with its usually great labellum, formed of one petal and two petaloid stamens, —with its singular pollen-masses, presently to be referred to, — with its column formed of seven cohering organs, of which three alone perform their proper function, namely, one anther and two generally confluent stigmas,—with the third stigma incapable of fertilisation and modified into the wonderful rostellum, — with three of the anthers no longer capable of producing pollen, but serv-

Consequently a section of the base of the labellum divides one lower petal, two petaloid anthers, portions of the two upper petals, and apparently of the two lower sepals and the two stigmas: altogether the section passes through the whole of or through portions of no less than seven or nine organs. The base of the labellum is here as complex an organ as the column of other Orchids.

ing either to protect the pollen of the fertile
anther, or to strengthen the column, or existing as mere rudiments, or entirely suppressed.
What an amount of modification, change of
function, cohesion, and abortion do we here
see! Yet hidden in that column, with its
surrounding petals and sepals, we know that
there are fifteen groups of vessels, arranged
three within three, in alternating order, which
probably have been preserved to the present
time from having been developed in each
flower at a very early period of growth, before
the shape or existence of this or that part
signified to the well-being of the plant.

Can we, in truth, feel satisfied by saying
that each Orchid was created, exactly as we
now see it, on a certain "ideal type;" that
the Omnipotent Creator, having fixed on one
plan for the whole Order, did not please to
depart from this plan; that He, therefore,
made the same organ to perform diverse functions—often of trifling importance compared
with their proper function—converted other
organs into mere purposeless rudiments, and
arranged all as if they had to stand separate,
and then made them cohere? Is it not a more

simple and intelligible view that all Orchids
owe what they have in common to descent
from some monocotyledonous plant, which,
like so many other plants of the same division,
possessed fifteen organs, arranged alternately
three within three in five whorls; and that
the now wonderfully changed structure of
the flower is due to a long course of slow
modification,—each modification having been
preserved which was useful to each plant,
during the incessant changes to which the
organic and the inorganic world has been
exposed?

On the gradation of Organs.—The rostellum,
the pollinia, the labellum, and, in a lesser
degree, the column, are the most remarkable
points in the structure of Orchids. Of the
two latter parts enough has been already said.
No organ like the rostellum exists in any
other flower. If the homologies of Orchids
had not been pretty well known, those who
believe in the separate creation of each being
might have advanced this as a case of a
perfectly new organ specially created, and
which could not have been developed by successive slow modifications of any pre-existing

part. But, as Robert Brown long ago remarked, it is not a new organ. It is impossible to look at the two groups of spiral
vessels (Fig. XXXII.) running from the mid-
ribs of the two lower sepals to the two,
sometimes quite distinct, lower stigmas, and
then to look at the third group of vessels
running from the mid-rib of the upper sepal
to the rostellum, which occupies exactly the
position of the third stigma, and doubt its
homological nature. There is every reason to
believe that the whole of this upper stigma,
and not merely a part, has been converted
into the rostellum; for there are plenty of
cases of two stigmas, but not one instance of
three stigmatic surfaces being present in those
Orchids which have a rostellum. On the
other hand, in Cypripedium and Apostasia
(the latter ranked by Brown in the Orchidean
order), which are destitute of a rostellum, the
stigmatic surface is trifid.

As we know only those plants which are
now living, it is impossible to follow all the
gradations by which the upper stigma has
been converted into the rostellum; but let us
see what are the facilities and indications of

such a change having been effected. The
change of function has not been so great as
it at first appears. The function of the
rostellum is to secrete a quantity of viscid
matter; it has lost that of fertility, but this
loss is so common with plants that it is hardly
worth notice. The stigmas of Orchids, as
well as of most other plants, secrete viscid
matter, the use of which in all cases is to
retain the pollen when brought by any means
to its surface. Now, if we look to one of the
simplest rostellums,—for instance, to that of
Cattleya or Epidendrum, — we find a thick
layer of viscid matter, not distinctly separated
from the viscid surface of the two confluent
stigmas : its action is simply to smear and
affix to the back of a retreating insect the
pollen-masses, which are thus dragged out of
the anther and transported to another flower,
where they are retained by the almost equally
viscid stigmatic surface. So that the office
of the rostellum is still to secure the pollen-
masses, but indirectly by means of their attachment to an insect's body.

The viscid matter of the rostellum and of
the stigma appears to have nearly the same

character : that of the rostellum generally has the peculiar property of quickly drying or setting hard; that of the stigma, when removed from the plant, apparently dried more quickly than gum-water of about equal tenacity. This tendency to dry is the more remarkable, as Gärtner * found that drops of the stigmatic secretion from Nicotiana did not dry in two months. The viscid matter of the rostellum in many Orchids when exposed to the air changes colour with remarkable quickness, and becomes brownish-purple ; and I have noticed a similar but slow change of colour in the viscid secretion of the stigmas of some Orchids, as of Cephalanthera grandiflora. When the viscid disc of an Orchis, as Bauer and Brown have also observed, is placed in water, minute particles are expelled with violence in a peculiar manner ; and I have observed exactly the same fact in the layer of viscid matter covering the stigmatic utriculi in an unopened flower of Mormodes ignea.

In order to compare the minute structure of the rostellum and stigma, I examined young flower-buds of Epidendrum cochleatum and

* 'Beiträge zur Kenntniss der Befruchtung,' 1844, s. 236.

floribundum, which, when mature, have a
simple rostellum. The posterior surface was
the same in both organs : the rostellum at
this early age consisted of a mass of nearly
orbicular cells, containing spheres of brown
matter, which resolve themselves into the
viscid matter : the stigma was covered with
a thinner layer of similar cells, and beneath
them were the coherent spindle-formed utriculi. The utriculi are believed to be connected with the penetration of the pollen-tubes;
and their absence in the rostellum probably
accounts for its infertility. As I do not find
the exterior layer of nearly orbicular cells
on the stigma in the bud-state (which apparently secrete the viscid matter), mentioned
by more experienced observers, I cannot help
feeling some doubt on the subject; though I
have no other reason to doubt my accuracy.
If the structure of the rostellum, in one of the
simplest Orchids, and of the stigma, be as I
have described them, their only difference is,
that in the rostellum the layer of cells, which
secretes the viscid matter, is thicker, and the
utriculi have disappeared.

Hence, during a course of slow modification, it is quite conceivable that the upper stigma, whilst still in some degree fertile or capable of penetration by the pollen-tubes, might secrete a superfluity of viscid matter; and that insects smeared with this might remove and transport the pollen-masses to the stigmas of other flowers. In this case an incipient rostellum would have been formed.

The following details on the rostellum and pollinia will have no interest to any one, unless he cares much about the structure of Orchids, or wishes to see how far very different states of the same organ can be graduated together within the limits of the same order. In the several Tribes, the rostellum presents a marvellous amount of diversity of structure; but most of the differences can be connected without very wide breaks. One of the most striking differences is, that either the whole anterior surface to some depth or only the central portion becomes viscid, in the latter case the surface retaining, as in Orchis, a membranous condition. But these two states so graduate into each other, that it is scarcely possible to draw any line of separation: thus, in Epipactis, the exterior surface undergoes a

vast change from its early cellular condition,
and becomes converted into a highly elastic
and tender membrane, which is in itself
slightly viscid, and allows the underlying
viscid matter readily to exude; yet it acts
as a membrane, with its under surface lined
with much more viscid matter : in Habenaria
chlorantha the exterior surface is highly
viscid, but still closely resembles, under the
microscope, the exterior membrane of Epipactis. Lastly, in some species of Oncidium,
&c., the viscid exterior surface differs, as far
as appearance under the microscope goes,
from the underlying viscid layer only in
colour; but it must have some essential difference; for I find that, until this very thin
exterior layer is disturbed, the underlying
matter remains viscid; whilst, after it has
been disturbed, it rapidly sets hard. The
gradation in the state of the surface of the
rostellum is not surprising, for in all cases in
the bud the surface is cellular; so that an
early condition has only to be more or less
perfectly retained.

The nature of the viscid matter differs
remarkably in different Orchids : in Listera

it sets hard almost instantaneously, more
quickly than plaster of Paris; in Malaxis
and Angræcum it remains fluid and viscid for
several days; but these two states pass into
each other by many gradations : in an Oncidium I have observed the viscid matter to
dry in a minute and a half; in some species
of Orchis in two or three minutes; in Epipactis in ten minutes; in Gymnadenia in two
hours; and in Habenaria in over twenty-four
hours. After the viscid matter of Listera has
set hard, neither water nor weak spirits of
wine has any effect on it; whereas that of
Habenaria bifolia, after being kept in spirits,
and after having been dried for several
months, when moistened was as sticky as
ever; the viscid matter in some species of
Orchis, when remoistened, presented an intermediate condition.

One of the most important differences in
the state of the rostellum is, whether or not
the pollinia are congenitally attached to it. I
do not allude to those cases in which the upper
surface of the rostellum becomes viscid, as in
Malaxis and some Epidendrums, and adheres
without mechanical aid to the pollen-masses;

for these cases present no difficulty, and can be graduated together. But I refer to the so-called congenital attachment of the pollinia by their caudicles. It is not, however, strictly correct to speak of congenital attachment, for the pollinia at an early period are invariably free, and become attached either earlier or later in different Orchids. No actual gradation is at present known in the process of attachment; but it can be shown to depend on very simple conditions and modifications. In the Epidendreæ the pollinia consist of a ball of waxy pollen, with a long caudicle (formed of elastic threads with adherent pollen-grains), which never becomes spontaneously attached to the rostellum. Cymbidium giganteum, on the other hand, has a congenitally attached caudicle, but its structure is identically the same, with this sole difference, that the elastic threads near its base adhere to, instead of simply lying on, the upper lip of the rostellum.

In an allied form, the Oncidium unguiculatum, I examined the development of the caudicles. At an early period the pollen-masses are enclosed in membranous cases;

which soon rupture at one point. At this early period, within the cleft of each pollen-mass, a layer of cellular matter may be detected, with the cells of rather large size, including remarkably opaque matter. The contained matter may be traced undergoing all the stages of development into the translucent substance forming the threads of the caudicles. As this change progresses, the cells themselves disappear. The threads at one end finally adhere to the waxy pollen, and at the other end, whilst still in a semi-developed state, they protrude through the small opening of the membranous case, and adhere to the rostellum, against which the anther is pressed. So that the adhesion of the caudicle to the back of the rostellum seems to depend solely on the early rupturing of the anther-case, and on a slight protrusion of the caudicles, before they have become fully developed and hardened.

In all Orchids a portion of the rostellum is, in fact, removed by insects when the pollinia are removed; for the viscid matter, though conveniently spoken of as a secretion, is a part of the rostellum in a modified condition.

But in those Orchids, which have their caudicles at an early period attached to the rostellum, a membranous or solid portion of its
exterior surface in an unmodified condition
is likewise removed. In the Vandeæ this
portion is sometimes of considerable size
(forming the disc and pedicel of the pollinium), and gives to their pollinia their most
remarkable character; but the differences in
the shape and size of the removed portions
of the rostellum can be finely graduated together, even within the Vandeæ; and still
more closely by commencing with the minute
oval atom of membrane to which the caudicle
of Orchis adheres, passing thence to that of
Habenaria bifolia, to that of H. chlorantha
with its drum-like pedicel, and thence through
many forms to the great disc and pedicel of
Catasetum.

In all the cases in which a portion of the
exterior surface of the rostellum is removed
attached to the caudicles, definite and often
complicated lines of separation are formed, or
are at least prepared by being weak, in order
to allow of the easy separation of the removeable portions. But the formation of the lines

of weakness does not differ much from certain
definite portions of the exterior surface of the
rostellum assuming a condition intermediate
between that of unaltered membrane and of
viscid matter, which has been already alluded
to. The actual separation of the lines depends
in many, perhaps in all, cases on the excitement of a touch ; and how this is effected is
at present inexplicable. But sensitiveness to
touch in the stigma (and the rostellum, as we
know, is a modified stigma), and indeed in all
the organs of vegetation, is not a very rare
attribute of many plants.

In Listera and Neottea, when the rostellum
is touched, even by a human hair, two points
rupture, and the included viscid matter is instantaneously expelled. Here we have a case
towards which as yet no gradation is known.
But Dr. Hooker has shown that the structure
of the rostellum is at first cellular (as in other
Orchids), and that the viscid matter originally
developed within these cells is contained,
apparently in a state of tension, in the loculi,
ready to be expelled as soon as the exterior
surface ruptures.

The last and conspicuous difference in the

state of the rostellum which I will mention is the existence, in many Ophreæ, of two widely-separated viscid discs, sometimes included in two separate pouches. Here it at first appears as if there were two rostella; but there is never more than one medial group of spiral vessels. In the Vandeæ we can see how a single viscid disc and single pedicel might become divided into two; for in some Stanhopeas the heart-shaped disc shows a trace of a tendency to division; and in Angræcum we have two distinct discs and two pedicels, either standing close together or removed a little way apart.

It might be thought that a similar gradation from a single rostellum into what appears like two distinct rostella was still more plainly shown in the Ophreæ; for we have the following series,—in Orchis pyramidalis a single disc enclosed in a single pouch; in Aceras two discs touching and affecting each other's shapes, but not actually joined; in Orchis latifolia and maculata two quite distinct discs with the pouch still showing plain traces of division; and, lastly, in Ophrys we have two perfectly distinct pouches, in-

cluding of course two perfectly distinct discs.
But this series does not indicate the former
steps by which a single rostellum has become
divided into two distinct organs ; but shows,
on the contrary, how the rostellum, after having been anciently divided into two organs,
has now in several cases been reunited into a
single organ.

This conclusion is founded on the nature of
the little medial crest (sometimes called the
rostellate process) between the bases of the
anther-cells (see Fig. I., B and D, p. 18).
In both divisions of the Ophreæ (those with
naked discs and those with discs enclosed in
a pouch), whenever the two discs come into
close juxtaposition, this medial crest or process
appears.* On the other hand, when the two
discs stand widely separate, the summit of the
rostellum between them is smooth, or nearly
smooth. In the Frog Orchis (Peristylus viridis) the overarching summit is bent like the

---

* Professor Babington ('Manual of British Botany,' 3rd edit.)
uses the existence of this "rostellate process" as a character to
separate Orchis, Gymnadenia, and Aceras from the other genera
of Ophreæ. The group of spiral vessels, properly belonging to
the rostellum, runs up, and even into, the base of this crest or
process.

roof of a house; and here we see the first stage of the formation of the folded crest. In Herminium, however, which has two separate and large discs, a crest, or solid ridge, is rather more plainly developed than might have been expected. In Gymnadenia conopsea, Orchis maculata, and others, the crest consists of a hood of thin membrane; in O. mascula the two sides of the hood have partly adhered; and in O. pyramidalis and in Aceras it has been converted into a solid ridge. These facts are intelligible only on the view, that, whilst the two discs were gradually, during a long line of generations, brought together, the intermediate portion or summit of the rostellum became more and more arched, until a folded crest, and finally a solid ridge, was formed.

Whether we compare together the state of the rostellum in the various Orchid-tribes, or compare the rostellum with the pistil and stigma of ordinary flowers, the differences are wonderfully great. A simple pistil of a common plant consists of a cylinder surmounted by a small viscid surface. Now, see what a contrast the rostellum of Catasetum, when dissected from the other elements of the column, presents; and as I traced all the vessels in this Orchid, the drawing may be trusted as approximately accurate. The whole organ has lost its normal function of fertility. Its shape is most singular, with its upper end

Fig. XXXIII.

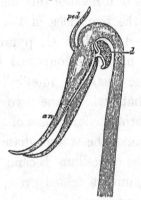

ROSTELLUM OF CATASETUM.

an, antennæ of rostellum.	*ped*, pedicel of rostellum, to which the pollen-masses are attached.
d, viscid disc.	

thickened, bent over and produced into two long tapering and sensitive antennæ, hollow within like adder's fangs. Behind and between the bases of these antennæ we see the large viscid disc, attached to the pedicel, which differs in structure from the underlying portion of the rostellum, and is separated from it by a layer of hyaline tissue, that spontaneously dissolves. The disc, attached to the surrounding parts by membrane which ruptures when excited by a touch, consists of strong upper tissue, with an underlying elastic cushion, coated with viscid matter; and this again in most Orchids is overlaid by a film of a different nature. What an amount of specialisation of parts do we here behold! Yet we have seen in the comparatively few Orchids described in this volume, so many and such plainly-marked gradations in the structure of this organ, and such plain facilities for the original conversion of the upper pistil into the rostellum, that it becomes far from incredible, if we had every Orchid which has ever existed throughout the world, that every gap in the existing chain, and every gap in many lost chains, would be amply filled up by a series of easy transitions.

We now come to the last great peculiarity in Orchids, namely, their pollen-masses. The anther opens early, and often deposits the naked masses of pollen on the back of the rostellum. This action is prefigured in Canna, a member of the family nearest allied to Orchids, in which the pollen is deposited on the pistil, close beneath the stigma. In the state of the pollen there is great diversity : in the anomalous Cypripedium single grains are embedded in a glutinous fluid ; in all other Orchids (except the degraded Cephalanthera) each grain consists of generally four united granules.* These compound grains are tied

---

* In several cases I have observed four tubes emitted from the four granules. In some semi-monstrous flowers of Malaxis paludosa, and of Aceras anthropophora, and in perfect flowers of Neottia nidus-avis, I have observed pollen-tubes emitted from the pollen-grains, whilst still within the anther and not in contact with the stigma. I have thought this worth mentioning, as R. Brown (in ' Linn. Transact.' vol. xvi. p. 729) states, apparently with some surprise, that the pollen-tubes were emitted from the pollen, whilst still within the anther, in a decaying flower of Asclepias. These cases show that the protruding tubes are, at least at first, formed at the expense of the contents of the pollen-grains.

Having alluded to the monstrous flowers of the Aceras, I will add that I examined several, always the lowest on the spike; in these the labellum was hardly developed, and was pressed close against the stigma. The rostellum was not developed, so that the pollinia had not viscid discs ; but the most curious feature was, that the two anther-cells had become, apparently in consequence of the position of the rudimentary labellum, widely separated, and were joined by a connective membrane, almost as broad as that of Habenaria chlorantha !

together by elastic threads, or are cemented together into so-called waxy masses by some unknown substance. The waxy masses thus formed are numerous in the Ophreæ; and graduate in the Epidendreæ and Vandeæ from eight, to four, to two, and, by the cohesion of the two, into a single mass. In the Epidendreæ we have both kinds of pollen within the same anther, namely, waxy masses, and caudicles consisting of elastic threads, with numerous compound grains adhering to them.

I can throw no light on the nature of the cohesion of the pollen in these waxy masses; when they are placed in water for three or four days, the compound grains readily fall apart; but the granules forming each grain still firmly cohere; so that the nature of the cohesion in the two cases is different. The elastic threads by which the packets of pollen are tied together in the Ophreæ, and which run far up inside the waxy masses of the Vandeæ, are also of a different nature; for they are acted on by chloroform, and by long immersion in spirits of wine; but these fluids have no particular action on the cohesion of the waxy masses. In several Epidendreæ

and Vandeæ the exterior pollen-grains of
these masses differ from the interior grains,
in being larger, and in having yellower and
much thicker walls. So that in the contents
of a single anther-cell we see a surprising
degree of differentiation of structure in the
pollen, namely, granules cohering by fours,
apparently due to their manner of early
development, and the compound grains partly
tied together by threads and partly cemented
together, with the exterior grains different
from the interior grains.

In the Vandeæ, the caudicle, composed of
fine coherent threads, is developed from the
semi-fluid contents of a layer of cellular membrane. As I find that chloroform has a peculiar and energetic action on the caudicles
of all Orchids, and likewise on the glutinous
matter enveloping the pollen-grains in Cypripedium, and which can easily be drawn
out into threads, one may suspect that in this
simpler Orchid we see the primordial condition
of the elastic threads by which the pollen-
grains in so many other more highly-developed
Orchids are tied together.*

* Auguste de Saint Hilaire states (' Lecons de Botanique,' &c.,

The caudicle, when largely developed and destitute of pollen-grains, is the most striking peculiarity of the pollinia.   In some Neotteæ, especially in Goodyera, we see it in a nascent condition, projecting just beyond the pollen-mass, with the threads only partially confluent. By tracing in the Vandeæ the gradation from the ordinary naked condition of the caudicle, through Lycaste in which it is almost naked, through Calanthe, to Cymbidium giganteum, in which it is covered with pollen-grains, it seems probable that its ordinary condition has been arrived at by the modification of a

1841, p. 447) that the elastic threads exist in the early bud, after the pollen-grains have been partly formed, as a thick creamy fluid. He adds that his observations on Ophrys apifera have shown him that this fluid is secreted by the rostellum, and is slowly forced drop by drop into the anther. Had not so eminent an authority made this statement I should not have noticed it. It is certainly erroneous. In buds of Epipactis latifolia I opened the anther, whilst perfectly closed and free from the rostellum, and found the pollen-grains united by elastic threads. Cephalanthera grandiflora has no rostellum to secrete the thick fluid, yet the pollen-grains are thus united. In a monstrous specimen of Orchis pyramidalis the auricles, or rudimentary anthers on each side of the proper anther, had become partly developed, and they stood quite on one side of the rostellum and stigma; yet I found in one of these auricles a distinct caudicle (which necessarily had no disc at its extremity), and this caudicle could not possibly have been secreted from the stigma. I could give additional evidence, but it would be superfluous.

pollinium like that of one of the Epidendreæ ;
namely, by the abortion of the pollen-grains
which primordially adhered to separate elastic
threads, and by the cohesion of these threads.

In the Ophreæ we have better evidence
than that offered by mere gradation, that the
long, rigid and naked caudicle has been thus
partly formed. I had often observed a cloudy
appearance in the middle of the translucent
caudicle ; and on carefully opening that of
Orchis pyramidalis, I found in several specimens in the centre, fully half-way down
between the packets of pollen and the viscid
disc, several pollen - grains (consisting, as
usual, of four united granules), lying quite
loose. These grains, from their embedded
position, could never by any possibility have
been left on the stigma of a flower, and were
absolutely useless. Those who can persuade
themselves that purposeless organs have been
specially created, will think little of this fact.
Those, on the contrary, who believe in the
slow modification of organic beings, will feel
no surprise that the process of change should
not always have been perfectly efficient,—
that, during and after the many inherited

stages of the abortion of the lower pollen-
grains and of the coherence of the elastic
threads, there should still exist a tendency
to the production of a few pollen-grains
where they were formerly developed, and
that these should consequently be left entan-
gled within the now coherent threads of the
caudicle. They will look at the little clouds
formed by the loose pollen-grains within the
caudicles of Orchis pyramidalis, as good
evidence that the pollen-mass of its parent-
form was originally like that of Epipactis or
Goodyera, and that the grains slowly disappeared from the lower parts of the mass,
leaving the elastic threads naked and ready to
cohere into a true caudicle.

As the caudicle, whether longer or shorter
in the several species, plays an important
part in the fertilisation of the flower, it apparently might have been formed from a nascent
condition, as in Epipactis, by the continued preservation of varying increments in its length,
each beneficial in relation to other changes in
the structure of the flower. But we may conclude from the facts given, that this has not been
the sole means,—that the caudicle owes much

of its length to the abortion of the lower pollen-
grains. That it has subsequently in some cases
been largely increased in length by natural
selection, is highly probable; for in Bonatea
speciosa the caudicle is actually more than
thrice as long as the elongated mass of pollen-
grains; and it is improbable that so lengthy a
mass of slightly cohering grains should have
existed, as an insect could not have safely
transported and applied to the stigma a pollen-
mass of this shape and size.

We have hitherto considered the gradations
in the state of the same organ. To any one
with much more knowledge than I possess,
it would be an interesting subject to trace,
in this great and closely-connected order, the
gradations, as far as possible, between the
several species and groups of species. To
make a perfect gradation, all the extinct forms
which have ever existed, along many lines of
descent converging to the common progenitor
of the order, would have to be called into life.
It is due to their absence, and to the consequent wide gaps in the series, that we are
enabled to divide the existing species into

definable groups, such as genera, families, and tribes. If there had been no extinction, there would still have been great lines, or branches, of special development, — the Vandeæ, for instance, would still, as a great body, have been distinguishable from the great body of the Ophreæ; but ancient and intermediate forms, very different probably from their present descendants, would have rendered it utterly impossible to separate by distinct characters the one great body from the other.

I will venture on only a few remarks. Cypripedium, in having three stigmas developed, and therefore in not having a rostellum, in having two fertile anthers with a large rudiment of a third, and in the state of its pollen, seems a remnant of the order whilst in a simpler condition. Apostasia is a related genus, placed by Brown amongst Orchids, but by Lindley in a small distinct family. These broken groups do not indicate to us the structure of the common parent-form of all Orchids, but they probably serve to show the state of the order in ancient times, when none of the forms had become so widely differentiated from each other and from other

plants, as are the existing Orchids, especially
the Vandeæ and Ophreæ; and when, consequently, the order made a nearer approach in
all its characters, than at present, to such
allied groups as the Marantaceæ.

With respect to other Orchids, we can see
that an ancient form, like one of the Pleuro-
thallidæ, some of which have waxy pollen-
masses with a minute caudicle, might give
rise, by the entire abortion of the caudicle,
to the Dendrobidæ, and by an increase of
the caudicle to the Epidendreæ. Cymbidium
shows us how simply a form like one of our
present Epidendreæ could be modified into
one of the Vandeæ. The Neotteæ stand in
nearly a similar relation to the Ophreæ, which
the Epidendreæ do to the Vandeæ. In certain
genera of the Neotteæ we have the pollen-
grains cemented into packets, tied together
by elastic threads, which project and form a
nascent caudicle. But this caudicle does not
protrude from the lower end of the pollinium as in the Ophreæ, nor does it always
protrude from the extreme upper end in the
Neotteæ; so that we can see that a transition
in this respect would be far from impossible.

In Spiranthes, the back of the rostellum, lined with viscid matter, is alone removed; the front part is membranous, and ruptures like the pouch-formed rostellum of the Ophreæ. An ancient form, combining most of the characters, but in a less developed state, of Goodyera, Epipactis, and Spiranthes, would, by further slight modifications, give rise to the whole tribe of the Ophreæ.

Hardly any question in Natural History is more vague and difficult to decide than what forms within any large group ought to be considered as the highest; * for all are excellently adapted to their conditions of life. If we look to traces of successive modification, with differentiation of parts and consequent complexity of structure, as the standard of comparison, the Ophreæ and Vandeæ will stand the highest. Are we to lay much stress on the size and beauty of the flower, and on the size of the whole plant? if so, the Vandeæ

---

* The fullest and the most able discussion on this difficult subject is by Professor H. G. Bronn in his ' Entwickelungs-Gesetze der Organischen Welt,' 1858. I have read the French translation published in 1861, Supplement, ' Comptes Rendus,' tom. ii. p. 520 et seq. This great work was crowned by the French Academy of Sciences.

are pre-eminent. They have, also, rather more complex pollinia, with the pollen-masses often reduced to two. The rostellum, on the other hand, in the Ophreæ, has apparently been more modified from its primordial stigmatic nature than in the Vandeæ. In all the Ophreæ the anthers of the inner whorl are almost entirely suppressed,—the auricles— mere rudiments of rudiments—being retained : these anthers have, therefore, undergone a greater amount of modification ; but can such suppression be considered as a sign of highness? I should doubt whether any member of the Orchidean order has been more profoundly modified in its whole structure than Bonatea speciosa, one of the Ophreæ. So again, within this tribe, nothing can be more perfect than the contrivances in Orchis pyramidalis for its fertilisation. Yet an ill-defined feeling tells me to rank the magnificent Vandeæ as the highest. When we look within this tribe at the elaborate mechanism of Catasetum for the ejection and transportal of the pollinia, with the sensitive rostellum so wonderfully modified, with the sexes borne on distinct plants, we may perhaps give to this genus the palm of victory.

A few miscellaneous points, which could not elsewhere have been conveniently introduced, deserve to be noticed. First, for the mechanism by which the pollinia in so many Orchids undergo a movement of depression, when removed from their places of attachment and exposed for a few seconds to the air. This is always due to the contraction of a portion, sometimes, as in Orchis, to an exceedingly minute portion, of the exterior surface of the rostellum, which has retained a membranous condition. This membrane, as we have seen, is likewise sensitive to a touch. In one Maxillaria the middle of the pedicel, and in Habenaria the whole drum-like pedicel, contracts. The point of contraction in all other cases seen by me is either close to the surface of attachment of the caudicle to the disc, or at the point where the pedicel is united to the disc; but both the disc and pedicel are parts of the exterior surface of the rostellum. In these remarks I do not refer to the movements simply due to the elasticity of the pollinia in the Vandeæ.

The long strap-formed disc of Gymnadenia
conopsea is well adapted to show the mechanism of the movement of depression.   The
whole pollinium, both in its upright and
depressed (but not closely depressed) position,
has been shown (p. 80) by Fig. X.   The disc,
highly magnified, in its uncontracted condition, is seen from above in the upper figure
here given, with the caudicle removed; and in
the lower figure we have a longitudinal section of the uncontracted
disc, together with the
base of the attached and
upright caudicle.   At the
broad end of the disc there
is a deep crescent-shaped
depression, bordered by
a slight ridge formed of
elongated cells.   The end of the caudicle is
attached to the steep sides of this depression
and ridge.   Now, when the disc is exposed
to the air for about thirty seconds, the ridge
contracts and sinks flat down; in sinking,
it drags with it the caudicle.   When placed
again in water the ridge rises, and when re-
exposed to the air it sinks, but each time with

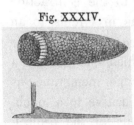

Fig. XXXIV.

Disc of Gymnadenia conopsea.

somewhat enfeebled power. With each sinking and rising of the caudicle, the whole
pollinium is depressed and elevated.

That the power of movement lies exclusively in the surface of the rostellum is well
shown by the saddle-shaped disc of Orchis
pyramidalis; for I removed under water the
attached caudicles, as well as the layer of
viscid matter from its under surface, and immediately that it was exposed to the air the
proper contraction ensued. The disc is here
formed of several layers of minute cells (and I
believe this to be the case with the disc of the
Gymnadenia), which are best seen in specimens kept in spirits of wine, for their contents
are thus rendered more opaque. The cells in
the flaps of the saddle are a little elongated.
As long as the saddle is kept damp its upper
surface is nearly flat, but when exposed to
the air (see Fig. III. E, p. 22) the surface
contracts immediately beneath the point of
attachment of the truncated end of each
caudicle, and becomes oblique ; and two
valleys are likewise formed in front of the
two caudicles. By this contraction the
caudicles are thrown down, almost in the

Q

same way as if trenches were dug in front of two upright poles, and the ground at the same time undermined beneath them. As far as I could perceive, an analogous contraction causes the depression of the pollinia in Orchis mascula.

Some pollinia which had been gummed on card for several months, when placed in water, underwent the movement of depression; and a fresh pollinium, when alternately damped and exposed to the air, can be made to rise and sink several times. Before I had ascertained these facts, which seem to show that the movement is hygrometric, I thought that it was a vital action, and tried vapour of chloroform and prussic acid, and immersion in laudanum; but these reagents did not check the movement. Nevertheless, there are considerable difficulties in understanding how the movement can be simply hygrometric. The flaps of the saddle in Orchis pyramidalis (see Fig. III. D, p. 22) curl completely inwards in nine seconds, which is surprisingly quick for the action of mere evaporation; and it is the under surface which curls inwards and ought to dry so quickly; but this cannot happen, as it is covered with a thick layer of viscid matter : the edges, however, of the saddle might become slightly dry in the nine seconds. When the saddle-formed disc is placed in spirits of wine it contracts energetically, and when placed in water opens again. This does not look as if the action was simply hydrometric. Whether the contraction is hygrometric, or is due to endosmose, or to some other unknown cause, the movements of depression in the pollinia thus produced are admirably regulated in each species, so that the pollen-masses, when transported by insects from flower to flower, should assume a position fitted to strike the stigmatic surface.

These movements would be quite useless, unless the pollinia first became attached to the insect in a uniform position relatively to the flower, so as to become after the movement of depression invariably directed towards the stigma ; and this necessitates that insects should be led to visit all the flowers of the same species in a uniform manner. Hence I must say a few words on the sepals and petals. Their primary function, no doubt, is to protect the organs of fructification in

Q 2

the bud. Even after the flower has fully
expanded, the upper sepal and two upper
petals often continue the same office. We
cannot doubt that this protection is of service,
when we see in Stelis the sepals so neatly
closing and reprotecting the flower after its
expansion; in Masdevallia the sepals soldered
together, with two little windows alone left
open; and when we see, in the open and
exposed flowers of the Bolbophyllum, that the
mouth of the stigmatic chamber after a time
closes. Analogous facts in Malaxis, Cephalanthera, &c., could be given. But the hood
formed by the upper sepal and two upper
petals, besides affording protection, evidently
forms a guide, compelling insects to visit the
flower in front. I do not believe that C. K.
Sprengel's view,* that the bright and conspicuous colour of the flower serves to attract
insects from a distance, is a fanciful notion;

---

* I am aware that this author's curious work, with its curious
title of 'Das Entdeckte Geheimniss der Natur,' has often been
spoken lightly of. No doubt he was an enthusiast, and probably
carried some of his ideas to an extreme length. I feel sure, from
my own observations, that his work contains a large body of
truth. Many years ago Robert Brown, to whose judgment all
botanists defer, spoke highly of it to me, and remarked that only
those who knew little of the subject would laugh at this work.

though some Orchids have singularly inconspicuous and greenish flowers,—perhaps in
order to escape some danger. Many of these
inconspicuous flowers are, however, strongly
scented, which would equally well serve to
attract insects.

But the labellum is by far the most important of the external envelopes of the flower.
It secretes, and often collects in a receptacle,
nectar; or is fleshy, and is furnished with
excrescences, which probably are attractive
to insects. Unless the flowers were by some
means rendered attractive, they would be
cursed with perpetual sterility. The labellum always stands in front of the rostellum,
and its outer portion often serves, as I have
seen, as a landing-place for the necessary
visitors: in Epipactis palustris this part is
flexible and elastic, and apparently compels
insects in retreating to brush against the
rostellum: in Cypripedium this end is folded
over like the end of a slipper, and compels
insects to insert their probosces over and near
the anthers. In the older flowers of Spiranthes
the labellum moves from the column, and
leaves a wider passage for the safe introduc-

tion of the pollinia, when attached to the proboscis of a bee. In certain exotic Orchids
the labellum suddenly moves and catches
insects as if in a box. In Mormodes ignea
it is perched on the summit of the column,
and here insects alight and touch the sensitive
hinge of the anther. The labellum is often
deeply channelled, or has guiding ridges, or is
pressed closely against the column, and in a
multitude of cases it approaches closely enough
to render the flower tubular. By these several
means insects are forced to brush against the
rostellum. We must not, however, suppose
that every detail of structure in the labellum
is of use : in some instances, as in Sarcanthus,
part of its extraordinary shape seems due to
its having been developed in the bud in close
apposition to the curiously shaped rostellum.

In Listera ovata the labellum stands far
from the column, but its base is narrow, so
that insects are led to stand exactly beneath
the middle of the rostellum : in other cases, as
in Stanhopea, Phalænopsis, &c., the labellum
is furnished with upturned basal lobes, which
manifestly act as lateral guides. In some
cases, as in Malaxis, the two upper petals are

curled backwards so as to be out of the way ;
in other cases, as in Acropera, Masdevallia,
and some Bolbophyllums, these upper petals
plainly serve as lateral guides, compelling
insects to visit the flower, or to insert their
probosces directly in front of the rostellum.
In other cases, wings from the margins of
the clinandrum, or from the sides of the
column, serve as lateral guides both in the
withdrawal of the pollinia, and in their subsequent insertion into the stigmatic cavity.
So that there can be no doubt that the petals
and sepals and rudimentary anthers do good
service in several ways, besides in affording
protection to the bud.

The final end of the whole flower, with all
its parts, is the production of seed; and these
are produced by Orchids in vast profusion.
Not that this is anything to boast of in the
order ; for the production of an almost infi-
nite number of eggs, or seeds, is undoubtedly
a sign of lowness of organisation. That a
plant, not an annual, should escape destruction
at some period of its life simply by the production of a vast number of seeds or seedlings,
shows a poverty of contrivance, or a want of

some fitting protection against some danger.
I was curious to estimate the number of seeds
produced by Orchids ; so I took a ripe capsule
of Cephalanthera grandiflora, and arranged
the seeds as equably as I could in a narrow
hillock, on a long ruled line ; and then counted
the seeds in a length, accurately measured,
of one-tenth of an inch. They were 83 in
number, and this would give for the whole
capsule 6020 seeds ; and for the four capsules
borne by the plant 24,000 seeds. Estimating
in the same manner the smaller seeds in
Orchis maculata, I found the number nearly
the same, viz. 6200 ; and, as I have often seen
above 30 capsules on the same plant, the
total amount will be 186,300,—a prodigious
number for one small plant to bear. As this
Orchid is perennial, and cannot in most places
be increasing in number, one seed alone of this
large number, once in every few years, produces a mature plant. I examined many seeds
of the Cephalanthera, and very few seemed
bad.

To give an idea what the above figures
really mean, I will briefly show the possible
rate of increase of O. maculata : an acre of

land would hold 174,240 plants, each having a space of six inches square, which is rather closer than they could flourish together; so that, allowing twelve thousand bad seeds, an acre would be thickly clothed by the progeny of a single plant. At the same rate of increase, the grandchildren would cover a space slightly exceeding the island of Anglesea; and the great grandchildren of a single plant would nearly (in the proportion of 47 to 50) clothe with one uniform green carpet the entire surface of the land throughout the globe.

What checks this unlimited multiplication cannot be told. The minute seeds within their light coats are well fitted for wide dissemination; and I have several times observed seedlings in my orchard, and in a newly-planted wood, which must have come from some little distance. Yet it is notorious that Orchids are sparingly distributed; for instance, this district is highly favourable to the order, for within a mile of my house nine genera, including thirteen species, grow; but of these one alone, Orchis morio, is sufficiently abundant to make a conspicuous feature in the vegetation; as is O. maculata

Q 3

in a lesser degree in open woodlands. Most
of the other species, though not deserving to
be called rare, are sparingly distributed; yet,
if their seeds or seedlings were not largely and
habitually destroyed, any one of them would,
as we have just seen, immediately cover the
whole land.

I have now nearly finished this too lengthy
volume. It has, I think, been shown that
Orchids exhibit an almost endless diversity
of beautiful adaptations. When this or that
part has been spoken of as contrived for
some special purpose, it must not be supposed
that it was originally always formed for this
sole purpose. The regular course of events
seems to be, that a part which originally
served for one purpose, by slow changes becomes adapted for widely different purposes.
To give an instance : in all the Ophreæ, the
long and nearly rigid caudicle manifestly
serves for the application of the pollen-grains
to the stigma, when the pollinium attached to
an insect is transported from flower to flower;
and the anther opens widely that the pollinium
may be easily withdrawn; but in the Bee
Ophrys, the caudicle, by a slight increase in

length, and decrease in thickness, and by the
anther opening a little more widely, becomes
specially adapted for the very different purpose
of self-fertilisation, through the combined aid
of the gravity of the pollen-mass and the
vibration of the flower. Every gradation
between these two states would be possible,
—of which we have seen partial proof in O.
arachnites.

Again the elasticity of the pedicel of the
pollinium in some Vandeæ is adapted to free
the pollen-masses out of their anther-cases;
but by further slight modifications, the elasticity of the pedicel becomes specially adapted
to shoot out the pollinia to a distance. The
great cavity in the labellum of many Vandeæ
serves to attract insects, but in Mormodes
ignea it is greatly reduced in size, and only
serves to keep the labellum in its proper position on the summit of the column. From the
analogy of many plants we may infer that a
long spur-like nectary is primarily adapted to
secrete and hold a store of nectar; but in
many Orchids it has so far lost this function,
as only to contain fluid between its two coats.
In those Orchids, in which the nectary con-

tains both free nectar and fluid in the intercellular spaces, we can see how a passage from
one state to the other could have been effected,
namely, by less and less nectar being secreted
from the inner membrane, and more and more
being retained within the intercellular spaces.
Other analogous cases could be given.

Although an organ may not have been
originally formed for some special purpose, if
it now serves for this end we are justified in
saying that it is specially contrived for it.
On the same principle, if a man were to make
a machine for some special purpose, but were
to use old wheels, springs, and pulleys, only
slightly altered, the whole machine, with all
its parts, might be said to be specially contrived for that purpose. Thus throughout
nature almost every part of each living being
has probably served, in a slightly modified
condition, for diverse purposes, and has acted
in the living machinery of many ancient and
distinct specific forms.

In my examination of Orchids, hardly any
fact has so much struck me as the endless
diversity of structure,—the prodigality of resources, — for gaining the very same end,

namely, the fertilisation of one flower by the pollen of another. The fact to a certain extent is intelligible on the principle of natural selection. As all the parts of a flower are co-ordinated, if slight variations in any one part are preserved from being beneficial to the plant, then the other parts will generally have to be modified in some corresponding manner. But certain parts may not vary at all, or may not vary in the simplest corresponding manner, and those variations, whatever their nature may be, which will bring all the parts into more perfect harmony with each other, will be seized on and preserved by natural selection.

To give a simple illustration : in many Orchids the ovarium (but sometimes the footstalk) becomes for a period twisted, causing the labellum to hang downwards, so that insects can easily visit the flower; but from slow changes in the form and position of the petals, or from new sorts of insects visiting the flower, it might become advantageous to the plant that the labellum should resume its normal upward position, as is actually the case with Malaxis paludosa; this change, it is obvious, might be simply effected by the continued

selection of varieties which had their ovarium
a little less twisted; but if the plant only
afforded varieties with the ovarium more
twisted, the same end could be attained by
their selection until the flower had turned
completely round on its axis: this seems to
have occurred with the Malaxis, for the labellum has acquired its present upward position,
and the ovarium is twisted to excess.

Again, we have seen that in most Vandeæ
there is a plain relation between the depth of
the stigmatic chamber and the length of the
pedicel, by which the pollen-masses are inserted; now if the chamber became slightly
less deep from any change in the form of the
column or any other unknown cause, the
shortening of the pedicel would be the simplest
corresponding change; but if the pedicel did
not happen to vary in length, any the slightest
tendency to an upward curvature from elasticity as in Phalænopsis, or to a backward
hygrometric movement as in one of the Maxillarias, would be preserved, and the tendency
would be continually augmented by selection;
thus the pedicel, as far as its action is concerned, would be modified in the same manner

as if it had been shortened. Such processes carried on during many thousand generations in various ways, with the several parts of the flower, would create an endless diversity of coadapted structures for the same general purpose. This view affords, I believe, the key which partly solves the problem of the vast diversity of structure adapted for closely analogous ends in many large groups of organic beings.

The more I study nature, the more I become impressed with ever-increasing force with the conclusion, that the contrivances and beautiful adaptations slowly acquired through each part occasionally varying in a slight degree but in many ways, with the preservation or natural selection of those variations which are beneficial to the organism under the complex and ever-varying conditions of life, transcend in an incomparable degree the contrivances and adaptations which the most fertile imagination of the most imaginative man could suggest with unlimited time at his disposal.

The use of each trifling detail of structure is far from a barren search to those who believe

in natural selection. When a naturalist casually takes up an organic being, and does not
study its whole life (imperfect though that
study will ever be), he naturally doubts whether each trifling point can be of any use, or
indeed whether it be due to any general law.
Some naturalists believe that numberless structures have been created for the sake of mere
variety and beauty, — much as a workman
would make a set of different patterns. I, for
one, have often and often doubted whether
this or that detail of structure could be of any
service; yet, if of no good, these structures
could not have been modelled by the natural
preservation of useful variations ; such details
could only be vaguely accounted for by the
direct action of the conditions of life, or the
mysterious laws of correlation of growth.

To give nearly all the instances of trifling
details of structure in the flowers of Orchids,
which are certainly of high importance, would
be to recapitulate a great portion of this
volume. But I will recall to the reader's
memory a few cases. I do not here refer to
the fundamental framework of the plant, such
as the remnants of the fifteen primary organs

arranged alternately in the five whorls; for
nearly all those who believe in the modification
of organic beings will admit that their presence
is due to inheritance from a remote parent-
form. A series of facts with respect to the
use of the variously shaped and placed petals
and sepals has just been enumerated. So,
again, the importance of the slight differences
in the shape of the caudicle of the pollinium
of the Bee Ophrys, compared with that of the
other species of the genus, has just been referred to : to this might be added the doubly-
bent caudicle of the Fly Ophrys : indeed, the
important relation of the length and shape of
the caudicle, with reference to the position of
the stigma, might be cited throughout whole
tribes. The solid projecting knob of the anther in Epipactis palustris, which does not include pollen, when moved by insects, liberates
the pollen-masses. In Cephalanthera grandiflora, the upright position of the flower, and
its almost closed condition, protect from dis-
turbance the slightly coherent pillars of pollen.
The length and elasticity of the filament of
the anther in certain species of Dendrobium
apparently serves for self-fertilisation, if in-

sects fail to transport the pollen-masses. The slight forward inclination of the crest of the rostellum in Listera prevents the anther-case being caught when the viscid matter explodes. The elasticity of the lip of the rostellum in Orchis causes it to spring up again when one pollen-mass is removed, thus keeping ready for action the second viscid disc, which otherwise would be wasted. The two nectar-secreting spots in the Frog Orchis, beneath the viscid discs at the base of the labellum, and the medial nectary in front of the stigma, are apparently necessary for the fertilisation of the flower. No one who had not studied Orchids would have suspected that these and many other small details of structure were of the highest importance to each species ; and that consequently, if the species were exposed to new conditions of life, and the structure of the several parts varied ever so little, such small details of structure might be modified by natural selection. These cases afford a good lesson of caution with respect to the importance, in other organic beings, of apparently trifling particulars of structure.

It may naturally be inquired, why do

Orchids exhibit so many perfect contrivances? From the observations of C. K. Sprengel, and from my own, I am sure that many other plants offer, in the means of fertilisation, analogous adaptations of high perfection; but it seems that they are really more numerous with Orchids than with most other plants. To a certain extent the inquiry can be answered. As each ovule requires at least one, probably more than one, pollen-grain,* and as the seeds produced by Orchids are so inordinately numerous, we can see that large masses of pollen would have to be left on the stigma of each flower. Even in the Neotteæ, which have granular pollen, with the grains tied together only by weak threads, I have observed that considerable masses of pollen are generally left on the stigmas. Hence we can perhaps understand the use of the grains cohering in large waxy masses, as they do in so many tribes, so as to prevent waste in the act of transportal. Most plants produce pollen enough to fertilise several flowers, even when each flower has several stigmas. But as the two

---

* Gartner, 'Beiträge zur Kenntniss der Befruchtung,' 1844, s. 135.

confluent stigmas of Orchids require so much pollen, its elaboration, if proportional in amount to that in most other plants, would have been extravagant in the highest degree, and exhausting to the individual. To save this waste and exhaustion, special and admirable contrivances were necessary for safely placing the pollen-masses on the stigma ; and thus we can partially understand why Orchids have been more highly endowed in this respect than most other plants.

The simple fact that many Vandeæ have only two pollen-masses, and that, from their coherence, some Malaxeæ have only a single pollen-mass, necessitates that extraordinary pains should have been taken in their fertilisation, otherwise these plants would. have been barren. The existence of a single pollen-mass, so that all the pollen-grains from one flower cannot possibly fertilise more than a single stigma, occurs, I believe, in no other plants. The case is partially analogous with that of seeds : many flowers produce a multitude of seeds, several produce a single seed ; very many flowers produce a countless number of pollen-grains ; some Orchid-flowers produce, as far as the power of fertilising other flowers is concerned, a single pollen-mass, though really consisting of a multitude of pollen-grains.

Notwithstanding that such care has been taken that the pollen of Orchids should not be wasted, we see that throughout the vast Orchidean order, — including, according to Lindley,* 433 genera, and probably about 6000 species,—the act of fertilisation is almost invariably left to insects. This assertion can hardly be considered rash, after the examination of so many British and exotic genera scattered through the main Tribes, which generally have a nearly uniform structure. In all plants in which insects play an important part in the act of fertilisation, there will be a good chance of pollen being carried from one flower to another. But in Orchids we have seen numerous adaptations,—such as the movements of the pollinia after their removal in order to acquire a proper position, —the slow movement of the labellum or rostellum to allow the entrance of the pollen-masses,—the separation of the sexes in some

* 'Gardener's Chronicle,' March 1, 1862, p. 192.

instances — which render it certain that in these cases the pollen of one flower or of one plant is habitually transported to another flower or plant. As this transportal increases the risk of loss, it necessitates and still further explains the extraordinary care bestowed on the contrivances for fertilisation.

Self-fertilisation is a rare event with Orchids. In Cephalanthera grandiflora it occurs, but in a very imperfect degree; and the early penetration of the stigma by the flower's own pollen-tubes seems to be fully as much determined by the support thus given to the pillars of pollen, as by the production of a small proportion of seed: certainly the fertilisation of this Orchid is aided by insects. In some species of Dendrobium self-fertilisation apparently occurs, but only if insects accidentally fail in removing the flower's own single pollen-mass. In Cypripedium, the Frog Orchis, and perhaps in a few other cases, it will depend on the manner (at present unknown) in which insects first insert their probosces by the one or the other entrance,—whether the flower's own pollen, or that of another flower, is habitually placed on the stigma; but in

these cases there will assuredly always be a
good chance of the stigma being fertilised by
pollen brought from another flower. In the
Bee Ophrys alone, as far as I have seen, there
are special and perfectly efficient contrivances
for self-fertilisation, combined, however, in
the most paradoxical manner, with manifest
adaptations for the occasional transport by
insects of the pollinia from one flower to
another, as in the other species of the same
genus.

Considering how precious the pollen of
Orchids evidently is, and what care has been
bestowed on its organisation and on the ac-
cessory parts ; — considering that the anther
always stands close behind or above the
stigma, self-fertilisation would have been an
incomparably safer process than the transportal of the pollen from flower to flower. It
is an astonishing fact that self-fertilisation
should not have been an habitual occurrence.
It apparently demonstrates to us that there
must be something injurious in the process.
Nature thus tells us, in the most emphatic
manner, that she abhors perpetual self-fertilisation. This conclusion seems to be of high

importance, and perhaps justifies the lengthy details given in this volume. For may we not further infer as probable, in accordance with the belief of the vast majority of the breeders of our domestic productions, that marriage between near relations is likewise in some way injurious,—that some unknown great good is derived from the union of individuals which have been kept distinct for many generations?

INDEX.